Crop Responses to Global Warming

Dinesh Chandra Uprety • V.R. Reddy

Crop Responses to Global Warming

Springer

Dinesh Chandra Uprety
Plant Physiology Division
Indian Agricultural Research Institute
New Delhi, India

V.R. Reddy
Change Laboratory (CSGCL)
Crop Systems and Global Change
Laboratory (CSGCL)
Beltsville, MD, USA

ISBN 978-981-10-2003-2 ISBN 978-981-10-2004-9 (eBook)
DOI 10.1007/978-981-10-2004-9

Library of Congress Control Number: 2016945745

Printed on acid-free paper

This Springer imprint is published by Springer Nature
The registered company is Springer Science+Business Media Singapore Pte Ltd.

Foreword

Global warming is the change in atmospheric temperature of the earth due to exponential rise of CO_2 and other greenhouse gases caused by anthropogenic and other activities. Rising temperatures are influencing various systems on the earth, including agriculture. In fact, global warming has a direct bearing on crops, food chains and production cycles in terms of marked changes on growth and yield processes of crops, insect and pest incidences, epidemics and increasing demands for irrigation resources. The impact assessment analysis of global warming has attracted the attention of the international scientific community. Recently India has set ambitious targets under its Intended National Determined Contributions (INDC) to the United Nations Framework Convention on Climate Change (UNFCCC) by extending its significant role in building additional carbon sinks. As evident in the context of India where an agro-based economy prevails, the direct impact of climate change jeopardizes the national food security. Global warming is taking a toll on our rich and diverse food systems. Reports indicate that the steep rise in temperature since 1980 has led to reduction in yields of staple crops offsetting gains even from improved farm practices, which has several implications for agriculture, crop yields and patterns in the long run. Therefore, futuristic research studies in these directions are needed. A comprehensive predictive model is yet to be established – a model that can improve and strengthen a plant's resilience to the stresses caused by increase in temperature.

The authors of the book Drs. Uprety and Reddy have given a comprehensive account of the historical rise of earth's temperature, its sources and sinks and more importantly the impacts on the production of various crop plants. They have also given illustrative descriptions of the impact of rise in temperature on the growth and various physiological processes to explain the response of crop plants to the change in temperature. The most interesting section of the book is the compilation of case studies on different crops such as wheat, rice, maize, soybean, cotton and brassica. The authors have made their best efforts to explain the impact of temperature changes on growth, yield and biochemical, physiological, morphological and agronomic systems of each species. In addition, mitigation technologies and future thrust sections for each crop have added value to this book at high stack. The book is timely and will help readers understand the complex issues of climate change implications for crops and, subsequently, food security. I wish them great success in every sphere of life and appreciate for taking us on this enlightening journey.

Director, G.B. Pant Institute of Himalayan
Environment and Development
Kosi-Katarmal, Almora, Uttarakhand, India

Dr. P.P. Dhyani

Preface

The monograph entitled "Crop Responses to Global Warming" describes the normal historical shifts in the earth's atmospheric temperature and weighs the evidence concerning anthropogenic induced changes in the level of temperature. There is not much study to explain the shift in temperature since the ice age, whereas the subsequent unprecedented increase in the earth's temperature after the preindustrial period has been possibly related to the anthropogenic activities. This rise in temperature has become a great threat to the productivity of crop plants. The increase in temperature has affected the crop calendar of tropical and temperate regions of the world by reducing the length of the effective growing season of crops. This monograph has given an overview of the global as well as Indian crop productivity in relation to the rise in the earth's surface temperature. The variability of past temperature changes and that of after the Industrial Revolution has been discussed in relation to the crop productivity and food security of the region. A chapter of this monograph is on the technologies to study the responses of crop plants to the elevated temperature. The innovative approaches have been described for the long-term experiments to investigate the lifetime responses of crop plants to the rise in temperature. Phytotrones, temperature gradient chambers, temperature gradient greenhouses (TGG), soil-plant-atmosphere research (SPAR) system, temperature-free air CO_2 enrichment technology (T-FACE), infrared warming technology (IRWT), free air temperature enrichment technologies (FATE), and soil warming systems (SWS) to simulate anthropogenic climate warming are described in this chapter.

The impact assessment analysis of rising temperature on crops such as wheat, rice, maize, soybean, cotton, and brassica are described, reviewed, and discussed in separate chapters as case studies. The responses of physiological processes and biochemical reactions to the elevated temperature in crop plants are described crop-wise. The effect of elevated temperature on the growth and development of crop plants has been discussed in relation to their phenological stages. The monograph also includes the impact of elevating temperature on crop/weed interaction, pest and diseases, and soil dynamics for each crop species independently. The mitigation technologies to counter the adverse effect of high-temperature stress are described for each crop according to their cultivation and climatic conditions. The future research strategies for each crop to meet the threat of elevating temperature on crop productivity and food security are described and discussed.

The monograph will assist in obtaining the detailed account on the historical analysis of the exponential rise in the temperature of the earth to the writers of climate change. The agricultural and environmental scientists will get the explanations of the phenological changes caused by rising atmospheric temperature and the physiological and biochemical characterization of these changes. The description of temperature enrichment technologies will help researchers and scientists to study the responses of biological materials to rising temperature. The monograph has first-hand knowledge on the mitigation technologies for ameliorating the global warming effects in crop plants. The detail description and suggestions on future strategies for each crop will be useful to the policy makers to change the farming policies to meet the challenges of global warming on crop plants. The monograph also describes the modification of crops and management systems to cope with changed temperatures which demonstrated the possibility of adoptions to high-temperature stress. The monograph will be the main text for teaching climate change, global warming, and environmental botany as no such book is currently available relating to the rising atmospheric temperature on crop plants. Therefore, the monograph will be highly useful for students of global climate change, environmental botany, and agricultural sciences, scientists, researchers, farmers, and policy makers.

New Delhi, India Dinesh Chandra Uprety
Beltsville, MD, USA V.R. Reddy

Acknowledgments

We acknowledge the work done by authors at Indian Agricultural Research Institute, New Delhi, and USDA-ARS, Crop Systems, and Global Change Research Lab at Beltsville, MD, USA, MD, USA, being the part of this book. The senior author acknowledges the contribution of his students, National Fellow research staff, and Emeritus Scientist laboratory researchers, without which the manuscript could not have been completed. He also thank his family members particularly his wife Kamla and son Pushpendra for their selfless help in completing the document. Authors are deeply indebted to Ms. Emily Warnack, Barnaby, Jin Young, and Dr. Jyostna Mura of USDA-ARS for editing and formatting the manuscript in a presentable shape. We acknowledge the help rendered by Mrs. Geetha Reddy not only during her personal visit to India but also by rendering the use of her immense knowledge of computer in correcting the manuscript. We thank Dr. Mamta Kapila in charge for Springer group for her help, encouragement, and guidance from time to time during the preparation of the document. Authors are also thankful to Ms. Raman Shukla of Springer group for her help in finalizing the administrative issues.

Dinesh Chandra Uprety
V.R. Reddy

Contents

Acronyms

APX	Ascorbate peroxidase
AQY	Apparent quantum yield
ATP	Adenosine triphosphate
BT	Basal temperature
CAT	Catalase
CGR	Crop growth rate
CER	CO_2 exchange rate
CIM	Composite interval mapping
DACT	Degrees above canopy threshold
DON	Deoxynivalenol
DT	Day temperature
DTI	Drought tolerance index
EIL	Economic injury levels
FACE	Free Air CO_2 Enrichment Technology
FATE	Free Air Temperature Enrichment Technology
Fv/Fm	Photochemical efficiency of PS ll
GB	Glycine betaine
GCM	General circulation model
GDD	Growing degree days
GLAI	Green leaf area index
GMT	Global mean temperature
GR	Glutathione reductase
IPCC	Intergovernmental Panel on Climate Change
IRRI	International Rice Research Institute, Philippines
IRTR	Infrared thermal radiometer
KGDD	Killing growing degree days
LEA	Late embryogenesis abundant protein
LTD	Leaf temperature difference
MDA	Malondialdehyde
MT	Million tonnes
NOAA	US National Oceanic and Atmospheric Administration
NT	Night temperature
NGDD	Normal growing degree days
OTC	Open-top chamber

PEP	Phosphoenolpyruvate
PID	Proportional integrative differential
PETM	Paleocene-Eocene Thermal Maximum
Pn	Net photosynthetic rate
PSl	Photosystem l
PSll	Photosystem ll
Q10	Temperature coefficient
QP	Photochemical quenching
QPSll	Quantum yield of PS ll electron transport
QTL	Quantitative trait loci
ROS	Reactive oxygen species
RLSU	Rubisco larger subunit
RSSU	Rubisco smaller subunit
RWC	Relative water content
SA	Salicylic acid
SFW	Shoot fresh weight
SDW	Shoot dry weight
SPAR	Soil-plant-atmosphere research system
SACC	Screen-aided CO_2 control system
SPS	Sucrose phosphate synthase
SUTl	Suc transporter gene
TFACE	Temperature free-air controlled enhancement
TGC	Temperature gradient chamber
TGG	Temperature gradient greenhouses
TBARS	Thiobarbituric acid reactive substances
TKW	Thermal kinetic window
T.min	Minimum temperature
T.opt	Optimum temperature
T.max	Maximum temperature
T.T.	Thermal time

Glossary of Terms

Abiotic Factors or processes of the nonliving environment (climate, geology, atmosphere, etc.).

Absolute growth rate The rate of increase in size of a growing plant (or part of it) in a given time under specific condition.

Acclimation Physiological adaptations to climate variations.

Adaptations Adjustment in natural and human system in response to actual or expected climatic stimuli or their effects which moderates harm or exploits beneficial opportunities.

Adaptive capacity The ability of a system to adjust to climate change to moderate potential damages, to take advantage of opportunities, or to cope with consequences.

Afforestation Direct human-induced conversion of land that has not been forested to forested land through planting.

Anthesis The period of opening of a flower, the period of pollination.

Antioxidant An organic compound that accepts free radicals and thus prevents the auto-oxidation of fats and oil.

Arid regions A land region of low rainfall (250 mm).

Anthropogenic Resulting from or produced by human beings.

Atmosphere A gaseous envelope surrounding the earth.

Barriers Any obstacle to reaching the goal, adaptation or mitigation potential that can be overcome by policy, program, and measure.

Biodiversity The total diversity of all organisms and ecosystems at various spatial scales (from genes to entire biomes).

Biomass The total mass of living organisms in a green area or volume.

Biome A biome is a major distinct regional element of the biosphere, typically consisting of several ecosystems (e.g., forests, rivers, ponds, and swamps within a region).

Biosphere The part of earth system comprising all ecosystems and living organisms in the atmosphere.

Biota All living organisms of an area, the flora and fauna considered as a unit.

C3 plants Plants that produce a three-carbon compound during photosynthesis, including most trees and agricultural crops such as rice, wheat, potatoes, and vegetables.

C4 plants Plants that produce a four-carbon compound during photosynthesis.

Carbon cycle The term used to describe the flow of carbon through the atmosphere, ocean, terrestrial biosphere, and lithosphere.

Carbon dioxide (CO_2) A naturally occurring gas, also a by-product of burning fossil fuels from carbon deposits such as oil, gas, and coal of burning biomass and of land use changes and other industrial processes.

Clean Development Mechanism (CDM) The CDM allows heat-trapping gas (greenhouse gas) emission reduction projects to take place in countries that have no emission targets under the United Nations Framework Convention on Climate Change (UNFCCC) Kyoto Protocol.

Climate Climate is defined as average weather over a period of time ranging from months to thousands or millions of years. The classical period for averaging these variables is 30 years as defined by WMO.

Climate change Climate change refers to a change in the state of the climate. Climate change may be due to natural internal processes or external forcing or to persistent human-induced changes in the composition of the atmosphere or in land use.

Climate model A numerical representation of the climate system based on the physical, chemical, and biological properties of its components and their interactions and feedback processes accounting for all or some of its known properties.

Crop growth rate It represents dry weight gained per unit area of crop in a unit time expressed as g $day^1 m^2$.

Deforestation Natural or human-induced processes that convert forests into non-forests.

Diurnal temperature range The difference between the maximum and minimum temperature during a 24-h period.

Drought Agricultural drought relates to moisture deficits in the topmost one meter or so of soil (root zone) that affect crops.

Ecosystem A system of living organisms interacting with each other and their physical environment.

Emissions Emissions of heat-trapping gases (greenhouse gases), greenhouse gas precursors, and aerosols associated with human activities including the burning of fossil fuels, deforestation, land use changes, livestock, fertilization, etc., that result in a net increase in emissions.

Energy balance The difference between total incoming and outgoing energy; if the balance is positive, warming occurs, and if it is negative, cooling occurs.

External forcing Forcing agent outside the climate system causing a change in climate system. Volcanic eruptions, solar variations, and human-induced changes in the composition of the atmosphere and land use changes are external forcing.

Food security A situation that exists when people have secure access to a sufficient amount of safe and nutritious food for normal growth, development, and an active and healthy life.

Fossil fuels Carbon-based fuels from fossil hydrocarbon deposits, including coal, peat, oil, and natural gas.

Genetically modified organism (GMO) An organism whose genes have been altered using genetic engineering techniques that transfer DNA molecules from different sources into an organism giving it modified or new genes.

Global surface temperature The global surface temperature is an estimate of the global mean surface air temperature.

Global warming It refers to the increase observed or projected, in global surface temperature, as one of the consequences of radiative forcing caused by human-induced emissions.

Greenhouse effect Greenhouse gases effectively absorb thermal infrared radiations by the earth's surface, by the atmosphere, and by clouds. Thus greenhouse gases trap heat within the surface.

Greenhouse gases These are the gaseous constituents of the atmosphere, both natural and anthropogenic, that absorb and emit radiation at specific wavelengths within the spectrum of thermal infrared radiation; water vapor (H_2O), carbon dioxide (CO_2), nitrous oxide (N_2O), methane (CH_4), ozone (O_3), and chlorofluorocarbon were included in this category.

Ground temperature The temperature of the ground near the surface (within the first 10 cm). It is often called soil temperature.

Ice age An ice age or glacial period is characterized by long-term reduction in temperature of the earth's climate resulting in the growth of ice sheets and mountain glaciers.

Impact assessment Identification and evaluation of the effect of climate change on natural and human systems.

Infrared radiation A form of energy not visible to the human eye but equally important as visible radiation on atmospheric processes. Often referred to as heat of radiation.

Invasive species Species aggressively expanding its range and population density in to a region in which it is not native, dominating native species.

Leaf area index The ratio between the total leaf surface area of a plant and the ground area covered by its leaves.

Mitigation Alleviation, lessening, reducing, and diminishing mitigation measures will aim to alleviate or offset the harm caused by a certain situation (climate change).

Pathogens Microorganisms such as bacteria or viruses that cause disease.

Phenology The study of natural phenomena that recur periodically (e.g., development stages, migration) and their relation to climate and seasonal changes.

Radiative forcing It is the change in the net downward minus upward irradiance in the troposphere due to change in an external driver of climate change.

Respiration The process whereby living organisms convert organic matter to carbon dioxide, releasing energy and consuming molecular oxygen.

Senescence Natural, pre-programmed aging and dying of a plant; this occurs after full maturity.

Sequestration Carbon storage in terrestrial or marine reservoir. Biological sequestration includes direct removal of CO_2 from the atmosphere through land use change, afforestation, reforestation, and carbon storage in land practices that enhance soil carbon in agriculture.

Sink A natural or technological process that removes carbon from the atmosphere and stores it.

Soil organic matter (SOM) Plant and animal residues, cells and tissues of soil organisms composed of carbon, and the substances the organisms synthesize.

Validate To establish or verify accuracy. For example, using measurements of temperature or precipitation to determine the accuracy of climate model results.

Vulnerability The degree to which the system is susceptible to, and unable to cope with, adverse effects of climate change including climate variability and extremes.

Water use efficiency Carbon gain in photosynthesis per unit water lost in evapotranspiration. Ratio of photosynthetic carbon gain per unit of transpirational water loss.

1 Introduction

Historical analysis showed temperature declined after A.D. 1000 until the late nineteenth century, followed by warming at a rate, unprecedented in the record. The change in temperature over the last 1000 years has been very small. Variability of temperature until the late nineteenth century can be accounted for by variation in the output of energy from the sun, explosive volcanic eruptions, and internal variability of the ocean-atmosphere system. The model estimate suggests additional temperature increases on the order of 1–2 °C by the end of the twenty-first century (IPCC 2001). However, the twentieth-century warming is inexplicable by these mechanisms. The increased level of greenhouse gases appeared to explain this unusual warming in the twentieth century. The rising concentration of greenhouse gases has resulted in a 0.76 °C increase in global surface temperature in the last 250 years. IPCC (2007) has now projected a temperature increase of 0.5–1.2 °C by 2020, 0.88–3.16 °C by 2050, and 1.56–5.44 °C by 2080 for Indian region, depending on the scenario for future development. Daily minimum temperatures are predicted to rise more rapidly than daily maximum temperatures. IPCC (2007) predicted that frost days will be lesser and extension of the growing season is likely to shift toward middle to high latitudes. Climate models predict that hot extremes, heat waves, and heavy precipitation events will become more frequent. The temperature increases are likely to be much higher in the winter season and precipitation is also likely to decrease. Some Indian studies also confirm a significant reduction in the agricultural production with an increase in the atmospheric temperature. There is a probability of 10–40 % loss in crop production in India with increase in temperature by 2080–2100. Winter crops such as wheat are especially vulnerable to high temperature during reproductive stages in India. Similarly, rice crop is also vulnerable to the rise in temperature causing about 5 % reduction in yield per 1 °C rise above 32 °C. Thus global warming is taking toll on our food. A steep rise in temperatures since 1980 has cut into yields of staple crops offsetting gains from genetically improved varieties and modified farm practices.

D.C. Uprety, V.R. Reddy, *Crop Responses to Global Warming*,
DOI 10.1007/978-981-10-2004-9_1

Table 1.1 NOAA data on July 2015

Global July 2015	Anomaly	Rank (out of 136 years)	Records
Land	$+0.96 \pm 0.18$		
	Warmest	6th (1998)	+1.11
	Coolest	131st (1884)	+0.68
Ocean	$+0.75 \pm 0.0$		
	Warmest	1st (2015)	+0.75
	Coolest	136th (1911)	−0.50
Land and Ocean	$+0.81 \pm 0.14$		
	Warmest	1st (2015)	+0.81
	Coolest	136th (1904, 1911)	−0.47

The world is in the midst of a very clear warming trend, where elevated carbon levels and natural temperature anomalies are causing more record-breaking hot years and fewer cold years. In fact 2014 was recently confirmed to be the hottest year ever recorded. While a record-breaking cold year has not seen for more than a century. The findings of the US National Oceanic and Atmospheric Administration (NOAA) showed a troubling trend as the planet continues to warm due to burning of fossil fuels. The July 2015 the average temperature across land and sea surfaces worldwide was 16.61 °C marking the hottest July ever in the 136 years period of record. The rise in the average temperature was 0.81 °C above the twentieth-century average of 15.8 °C surpassing the previous record set in 1998 by 0.08 °C. This was also the all-time highest monthly temperature in the 1880–2015 record, according to the monthly (July 2015) report of NOAA (Table 1.1).

The accumulation of greenhouse gases and their continued emission will possibly increase the annual temperatures by 2.5–4.3 °C in the crop-growing regions of the world by 2080–2099 (Le Quere et al. 2009). Growing season temperatures may rise more than the annual averages with reduced precipitation. Future agricultural production will encounter multifaceted challenges from such increasing trend of global temperature. This rise of temperature in the tropics and subtropics may exceed even the most extreme seasonal temperature (Battisib and Naylor 2009). The Middle East and Africa have started reporting crop yield decline. India produced 1 million metric tonnes less wheat in 2014 than it did in 2013. Recent studies have indicated the possibility of 10–40 % loss in crop productivity in India with increase in temperature by 2080–2100 (Parry et al. 2004). Crop simulation models indicate a possibility of 4–5 million tonnes loss in wheat production for every 1 °C rise in temperature throughout the growing season (Aggarawal 2008). Food production needs to grow 66 % more by 2050 to meet the projected demand from an anticipated population of more than 9 billion people (FAO 2009).

However, according to the latest IPCC assessment report, adaptations of agriculture will result in increased yields of cereal crops (maize, wheat, and rice) in the mid- and high-altitude regions with modest increases in temperature (Easterling et al. 2007). The global analysis of crop yield from 1981 to 2002 indicated a negative response of wheat, maize, and barley yields due to rising temperature costing an estimated $5 billion per year (Lobell and Field 2007). The response of maize and soybean to temperature was nonlinear, and the decline in yield above optimum temperature is significantly steeper ranging between 30 and 46 %. According to Lobell et al. (2008), wheat and rice in Southeast Asia and maize in southern Africa were most likely to be negatively affected by rising temperature in the absence of adaptation strategies.

This exponential rise in temperature is considered to be the potential threat to the productivity of agricultural crops. The male gametophyte development in wheat and rice is vulnerable to a high temperature causing reduction in their grain yield. High leaf temperature is also a consequence of drought which results in the loss of crop's ability for transpiration cooling thereby adversely affecting the yield. Such an effect was demonstrated in cotton by Radin et al. (1994). Negative outcome of high temperature was visible in the productivity of millets, groundnuts, and rape seed mustards in southern Asia, sorghum in Sahel, and maize in southern Africa. The rising temperature during crop-growing seasons is also a serious threat for food security.

The magnitude of increase in temperature has been reported to be higher at high latitude as compared to low latitude. The projected warming in the twenty-first century is expected to be greatest over land and at most high in Northern latitudes and least over the Southern Ocean and part of North Atlantic Ocean. Global warming may be beneficial for alpine regions, mainly by expanding arable lands and extending crop growth duration but harmful for subtropical and tropical regions. High temperature may lead to spatial and temporal loss of crop biodiversity and geographical shifting of temperate crops from their traditional areas of cultivation to upper latitude for acquiring optimum degree days to flowering and fruiting. Rising temperature may reduce the resource utilization efficiency of crops mainly due to imbalance in source sink relationship resulting in to decrease in the biomass and grain/seed production. It may also cause the spread of existing disease and insect pests and emergence of new biotypes and diseases. There is also possibility of greater crop/weed competition for their growth and productivity.

Warming is expected to extend a northward expansion of suitable cropping areas and a reduction of the growing period of determinate crops (e.g., cereals), but an increase in indeterminate crops (e.g., root crops). Cereals, oilseed, and protein crops including pulses are generally determinate species, and their duration to maturity depends on the temperature of the growing areas and in many cases day length. A temperature increase will, therefore, shorten the length of the growing period and reduce yields, if management is not altered (Porter and Gawith 1999). Climate change may produce positive effects on agriculture in the northern areas through the introduction of new suitable crop species and varieties, which produce greater yields. The expansion of growing areas for the cultivation of such crop species will be advantageous in exceeding their production. However, the disadvantages of the warming may be an increase in the need for plant protection, the risk of nutrient leaching, and the turnover of soil organic matter. In the southern areas the possible increase in water shortage and extreme weather events may cause lower harvestable yield variability and a reduction in suitable areas for traditional crops.

Olesen and Bindi (2002), while describing the consequences of climate change for European agricultural productivity, explained that at middle and higher latitudes of Europe, global warming will extend the length of the potential growing season, allowing early planting in the spring followed by earlier maturation and harvesting of crop plants. Less severe winters will allow the growing of more productive cultivars of winter annual and perennial crops. This is important for C_4 species, since the pyruvate phosphate dikinase is sensitive to low temperature. This enzyme plays a key role in the regeneration of phosphoenol pyruvate, the acceptor of CO_2 in C_4 species. The increased temperature increases the respiration in crops resulting in the reduced growth of crop plants in lower latitude area of Europe. Another important effect of high temperature is accelerated development, hastened maturation, and reduced yield of determinate crops. The cropping area may expand northward in countries such as Finland and Russia. Increase in annual mean temperature will make central Europe suitable for warmer-season crops like grain maize and grapevine. Global warming will extend the length of the potential growing season, allowing earlier planting of crops in the spring, earlier maturation, and harvesting in the middle and higher latitudes of Europe. Carter et al. (1996) found a northward shift of areas suitable for spring cereals of 120–150 km per 1 °C increase in annual mean temperature. Spatial shifts northward and into central Europe have also been estimated for warmer-season crops like grain maize and grapevine (Kenny and Harrison 1992; Kenny et al. 1993). Rise in temperature in warmer, lower latitude regions of Europe would increase the respiration rates of crop plants, resulting in less than optimal conditions for net crop growth. An important effect of high temperature is accelerated development, causes hastened maturation of determinate crops results a grave reduction in their grain yield and productivity (Rötter and van de Geijn 1999).

References

Aggarawal PK (2008) Global climate change and Indian agriculture: impacts, adaptation and mitigation. Indian J Agric Sci 78(10):911–919

Battisib DS, Naylor RL (2009) Historical warnings of future food insecurity with unprecedented seasonal heat. Science 323:240–244

Carter TR, Saarikko RA, Niemi KJ (1996) Assessing the risks and uncertainties of regional crop potential under a changing climate in Finland. Agric Food Sci Finl 3:329–349

Easterling W, Aggarwal P, Batima P et al (2007) Food, fibre and forest products. In: Oarry ML et al (eds) Climate change 2007: impacts, adaptation and vulnerability. Cambridge University Press, Cambridge, pp 273–313

FAO (Food and Agriculture Organization) (2009) How to feed the world in 2050. http://www.fao.org/fileadmin/templates/wsfs/docs/expert_paper/How_to_Feed_the_World_in_2050.pdf. Accesses 24 Feb 2016

IPCC (Intergovernmental Panel on Climate Change) (2001) Climate change 2001: the scientific basis- contribution of working group l to the third assessment report of the IPCC, Cambridge

IPCC (Intergovernmental Panel on Climate Change) (2007) Summary for policy makers. In: Parry ML (ed) Climate change 2007: impacts, adaptation and vulnerability, contribution of working group II to the fourth assessment report of the Intergovernmental Panel on Climate Change. Cambridge University Press, Cambridge

Kenny GJ, Harrison PA (1992) The effects of climate variability and change on grape suitability in Europe. J Wine Res 3:163–183

Kenny GJ, Harrison PA, Olesen JE et al (1993) The effects of climate change on land suitability of grain maize, winter wheat and cauliflower in Europe. Eur J Agron 2:325–338

Le Quere C, Raupach MR et al (2009) Trends in the sources and sinks of carbon dioxide. Nat Geosci 2:831–836

Lobell DB, Field CB (2007) Global scale climate-crop yield relationships and the impact of recent past. Environ Res Lett 2(014002):7p

Lobell DB, Burkee MB, Tebaldi C et al (2008) Prioritizing climate change adaptation and needs for food security in 2030. Science 319:607–610

Olesen JE, Bindi M (2002) Conséquences of climate change for European agricultural productivity, land use and policy. Eur J Agron 16:239–262

Parry ML, Rosenzweig C, Iglesias A et al (2004) Effects of climate change on global food production under SRES emissions and socio-economic scenarios. Global Environ Change 14:53–67

Porter JR, Gawith M (1999) Temperatures and the growth and development of wheat: a review. Eur J Agron 10:23–36

Radin JW, Lu ZM, Percy RG et al (1994) Genetic variation for stomatal conductance in Pima cotton and its relation to improvements for heat adaptation. Proc Natl Acad Sci 91:7217–7221

Rötter R, van de Geijn SC (1999) Climate change effects on plant growth, crop yield and livestock. Climate Change 43:651–681

Problems and Prospects of Crops with Changing Temperature

2

2.1 Soil

As soil organic matter decomposes, CO_2 is released back into the atmosphere, which could further exacerbate additional global warming (Carney et al. 2007). The decomposition is faster in the summer and slow in the winter. Soil has a relatively large capacity to store energy, and therefore, soil temperature does not change rapidly. The temperature manipulation experiments (Rustad et al. 2001) have shown that the rate of soil respiration increases with warmer temperatures and demonstrated a significant positive relationship between temperature and soil respiration. The magnitude of the response of soil respiration to warming was greater in colder and high-altitude ecosystems. It was also suggested by them that some of the ecosystems of high altitudes have changed from a net carbon sink to a net carbon source. The soil decomposition is faster in the summer and slower in winter. This relationship between air temperature, soil moisture, and soil decomposition is important and predictable for the climate change-induced dynamic shifts in temperature. However, soil temperature is linked to canopy temperature via energy exchange within the canopy as mediated by resistances between canopy, canopy air, and soil surface. Besides convection, the soil can also gain energy directly from the leaves by reemitted thermal radiations. Thus soil temperature follows canopy temperature considering other variables (Luo et al. 2010). According to Wall et al. (2013) global warming will likely affect carbon cycles in agricultural soils. The infrared heater experiments suggest that soil warming only affects soil processes with little effects on plant processes like photosynthesis and transpiration in the upper canopy. Infrared warming increased the soil temperature and decreased volumetric soil water. The effects of humidity and temperature on stomatal conductance are secondary. IR warming initially increased soil CO_2 efflux following soil hydration but as the soil surface dried, soil CO_2 efflux decreased even under warmer temperature. The supplemental irrigation can minimize this artifact by increasing the vapor pressure gradient. Thus, the regions of the earth that contains high soil

D.C. Uprety, V.R. Reddy, *Crop Responses to Global Warming*,
DOI 10.1007/978-981-10-2004-9_2

carbon substrate, with supplemental irrigation in the future, will likely exhibit an increase in soil CO_2 efflux. The CO_2 efflux is likely to decrease for many semiarid desert regions that contain low soil carbon as the earth becomes warmer. Many studies (Wall et al. 2013) demonstrated that the warming of soil would result in a greater storage of carbon in plant biomass and increased flux of carbon to soils in the form of leaf and root litter. This could at least partially offset the warming-induced increase in carbon flux from soils to the atmosphere via soil respiration. The negative response to increasing temperature is the concomitant decline in soil moisture associated with higher temperature.

2.2 Growth and Development of Crop Plants

High-temperature stress effects from germination to plant growth and development and controls various physiological processes such as the photosynthetic functions and reproductive processes, namely, flowering, pollination, and seed set. For example, following phenological stages in wheat is controlled by temperature:

1. Germination, i.e., seeding to seedling emergence regulated by low temperature
2. Canopy development, i.e., emergence of first spikelet initiation and double-ridge stage
3. Spikelet production: first spikelet initiation to terminal spikelet formation
4. Spikelet development: terminal spikelet formation to anthesis
5. Grain development: anthesis to maturity

A number of leaf and tiller primordial are determined before spikelet initiation, but their subsequent growth and development are controlled by the temperature and day length. Similarly floret number within each spikelet is established by anthesis at which time the potential grain number per spike is determined. Other factors, that control the high-temperature effects on plant growth and development and help in grain development and yield enhancement, are the increase in the rooting depth to explore water from the deep soil, slower rate of leaf emergence, canopy development, and evapotranspiration at baseline temperature. The early stomatal closure due to high temperature is predicted to reduce yield and determines variable responses among cultivars. However, such response of stomata to high temperature did ensure that the crop avoided a high-temperature stress that reduces the yield, allowing stomata to remain open even at very low water potential via osmotic adjustment.

Among the extreme temperatures, the high-temperature stress was more frequently recorded compared to cold-temperature stress. The increase in temperature lengthens the effective growing season in the northern temperate region and causes a poleward shift of thermal limits to agriculture. This poleward shift will help for sustainable productivity of crops like rice, which is of tropical origin and adaptations. However, rain-fed wheat depends on the availability of soil moisture after monsoon rain recedes in September. High maximum and minimum temperature in

September (about 34/20 °C) adversely affects seeding establishment, accelerates early vegetative development, and reduces canopy cover, tiller initiation, spike size, and yield. Hence sowing is typically delayed until after mid-October when seed beds had cooled, though much of residual soil moisture may have lost. High temperature in the second half of February (25/10 °C), March (30/13 °C), and April (30/20 °C) reduces the number of viable florets and grain filling duration. High-temperature stress, particularly reduces yield of wheat sown in December/January, which is prevalent in some regions because of the multiple cropping system. The higher temperature in annual crops can drive shorter life cycles, resulting in less seasonal photosynthesis, shorter reproductive phase, and thus lower yield. Vegetative development is accelerated with the rise in temperature in cereals but the shorter grain filling period drastically affected the grain yield. According to Craufurd and Wheeler (2009), the interaction of temperature and photoperiod is highly important in determining the flowering time. High-temperature stress that affects any of the reproductive processes, including pollen viability, female gametogenesis, pollen-pistil interaction, fertilization, and grain formation, severely reduces the grain yield (Hedhly et al. 2008).

Response of crop plant species to the climate change variables

Species	Comments	Reference
Wheat	No change	Krenzer and Moss (1975)
Wheat	Shorter phyllochron interval in summer than winter	Gifford (1977)
Wheat	No change	Kendall et al. (1985)
Wheat	Early anthesis,accelerated grain growth,greater respiratory loss	Uprety et al. (2009)
	Lower grain yield	
Soybean	Little effect on phenology	Allen et al. (1988)
Soybean	Shorter plastochron interval	Baker et al. (1989)
Brassica	Early flowering	Uprety et al. (1995)
Cowpea	Early flowering	Bhattacharya et al. (1985)
Rice	Early flowering, low flag leaf area & number	Imai et al. (1985)
Rice	Early flowering,reduced photosynthesis &low yield	Uprety et al. (2000)
Sunflower	Slightly delayed flowering	Hesketh and Hellmers (1973)
Sorghum	Late flowering	Hesketh and Hellmers (1973)
Cotton	Delayed flowering	Hesketh and Hellmers (1973)
Maize	Slightly late flowering	Hesketh and Hellmers (1973)

The growth and development in crop plants show critical optimal temperature for each process, whereby process rates increase over a range but thereafter flatten and decrease. For example, the light-saturated photosynthesis rate of C_3 crops such as wheat and rice is at a maximum for temperatures between 20 and 32 °C; total crop respiration (growth and maintenance) shows a steep nonlinear increase with temperatures from 15 to 40 °C followed by a rapid and nearly linear decline. Any constraint in photosynthesis can limit plant growth at higher temperature. Wise et al. (2004) reported that photochemical reactions in thylakoid lamellae including photosynthetic electron flow and carbon metabolism in the stroma of chloroplast are the primary sites of injury of high temperature. According to Camejo et al. (2005), PS II is greatly reduced under high temperature which may be due to the temperature-induced injury on thylakoid membrane, wherein the PSII system is located. On the other hand, high temperature may lead to the dissociation of oxygen-evolving complex, resulting in to imbalance between the electron flow from oxygen-evolving complex toward the acceptor side of PS II in the direction of the PS I reaction center (De Ronde et al. 2004). Sujatha et al. (2008) linked the reduction in the rate of photosynthesis to the increase in temperature with the decrease in the activity of sucrose phosphate synthase enzyme and increase in the accumulation of soluble sugars and starch in rice. They suggested that the upregulation of sucrose phosphate synthase will be useful for the growth and productivity of rice. The impact on photosynthesis is being regulated by the differential changes on the solubility of CO_2 and O_2 due to higher temperature.

Porter and Semenov (2005) demonstrated that temperature thresholds are well defined, above which the formation of reproductive sinks such as seeds and fruits are adversely affected Table 2.1. High temperature during reproductive development is particularly injurious if they occur just before or during anthesis resulting in lower seed set in crop species such as *common bean* (*Phaseolus vulgaris*) (Prasad et al. 2002), peanut (*Arachis hypogea*) (Prasad et al. 2003), cow pea (*Vigna unguiculata*) (Hall 1992), tomato (*Solanum lycopersicum*) (Reddy et al. 2000), wheat (*Triticum aestivum*) (Saini and Aspinall 1982), and rice (*Oryza sativa*) (Satake and Yoshida 1978). The critical temperature for various processes in different species of crop plants has been determined, for example, peanut (*Arachis hypogea*) experiences heat stress at temperature above 40 °C, the percentage fruit set fell from 50 % of flowers to zero, and the decline rate was linear. Similarly maize showed reduced pollen viability above 36 °C, rice grain sterility is brought on by temperatures in the mid-30 °C, and similar temperatures can reverse the vernalizing effects of cold temperatures in wheat. Crop modeling can predict the timing of the stages when crops are sensitive to threshold temperatures; experimental studies can provide the quantitative responses that will permit the modeling effort to progress.

Plant physiological and growth processes are regulated more directly by the temperatures of the tissues than by air temperatures which can differ from plant canopy temperatures by several degrees (Lecain et al. 2015). Moreover, the degrees of global warming itself are more driven by the reductions in the rate of radiant energy loss from crop canopies and other surfaces that are being partially closed by increases in atmospheric CO_2 and water vapor than by the consequent warming of air temperature. Porter and Semenov (2005) explained that the variability of temperature affects the physiological processes related to growth and productivity. Some crop processes, mostly related to growth, such as photosynthesis and respiration, show continuous and mainly nonlinear changes in their rates as temperature increases. The light-saturated photosynthesis rate of C_3 crops such as wheat and rice is at a maximum for temperature from about 20 to 32 °C; total crop respiration shows a steep nonlinear increase in the temperature from 15 to 40 °C followed by a rapid and linear decline. Rapid changes in enzymatic reactions caused by differential thermosensitivity of various enzymes can deplete or result in the accumulation of metabolites. The short-term effects involving altered gene expression, such as heat shock protein synthesis, may also help in countering the heat stress effects in crop plants. The longer-term effects include changes in the rate of photosynthesis and electron transport per unit leaf area and impaired cell anaplerotic carbon metabolism, sucrose synthesis and carbon-nitrogen partitioning within and between organs. Altered carbon availability brought about by these events will affect uptake, transport, and assimilation of other nutrients and disturb lipid metabolism and injured cell membranes (Maheswari et al. 1999). Crops vary in their ability to survive high- and low-temperature stresses. However, the determination of crop plants to survive or adapt to the changes in temperature needs to be understood in various crop species individually. This is extremely important to develop cultivars to survive and sustain such temperature extremes. This effort has brought breeders and crop physiologists together to assess the yield stability and crop performance in the field to meet the challenges of heat and drought stresses.

2.3 Photosynthesis and High-Temperature Stress

High temperature is the most prominent driver that determines the crop growth and productivity. Thermal stability of the photosynthetic apparatus differed markedly between species from temperate and tropical environments. Temperate cereals usually lack the thermotolerance like most tropical species and frequently suffer greater injury during stress, High temperature also damages photosynthesis before other activities in sensitive species. Photosynthesis and Hill reaction declined earlier than other physiological processes in wheat leaves (Al Khatib and Paulson 1999). Sensitivity of photosynthesis to high temperature in temperate species was attributed to vulnerability of PSll in thylakoid membrane (Santarius 1973). The threshold temperature for denaturation of PSll in *Triticum durum* ranged from 35 to 41 °C (Rekika et al. 1997). Temperature up to 35 °C during maturation of wheat accelerated loss of chloroplast integrity and PSll activity but had little effect on PSl (Xu et al. 1995). Stromal enzymes are more stable to high temperatures than is PSll. The most severe injury by high temperature to CO_2-limited photosynthesis than to CO_2-saturated photosynthesis observed in wheat in comparison to rice. This variation was attributed to the variation in the kinetic properties of carboxylase relative to the oxygenase activities of Rubisco between wheat and rice with the rise in temperature. Changes in relative solubility of CO_2 and O_2 with temperature might have differentially affected photosynthesis and photorespiration. Rubisco activase, which is highly vulnerable to temperature changes, might be responsible for the variation in the response of wheat and rice photosynthesis. The C_4 photosynthetic pathway, which functions as a CO_2-concentrating mechanism, markedly improved the assimilation at high temperature. The C_4 pathway appears to be inadequate to accord adaptation to warmer temperature. Thus the noncyclic photophosphorylation appears to be a necessity to meet the challenge of higher temperature (Khatib and Paulsen 1999).

High photosynthesis at high temperature does not support high rate of dry matter accumulation. Photosynthesis rates were found to be temperature sensitive in other crops; however, no measurable change was observed in the photosynthesis rate of wheat leaves between 15 and 35 °C (Bagga and Rawson 1977). There was little temperature effect on photosynthesis rate of rice leaves from 20 to 40 °C (Egeh et al. 1994). However, elevated CO_2-induced increase in photosynthesis was more in warmer environment (Gifford 1989). The effects of warmer temperature on photosynthesis will be one of the most important determinants of crop yield. The Rubisco activity is directly affected by temperature. The Rubisco is quite thermally stable and the carboxylation activity continues to increase beyond 50 °C. However, it loses its discrimination power resulting in to an increase in the solubility of O_2 compared to CO_2 and increasing the photorespiration and inhibiting the carbon assimilation in C_3 plants. According to June et al. (2004), the rate of regeneration of CO_2 acceptor RuBP is more sensitive to temperature. Additionally RuBP regeneration limitation increases at high temperature (Sage and Kubian 2007). Thus improving the temperature response of photosynthesis by manipulating RUBP regeneration appears to be a promising option for adaptation/mitigation of high-temperature stress. The recent discovery by Salvucci (2008) on the association of Rubisco activase with the chloroplast GroEL suggests that this protein may provide mechanism to protect Rubisco activase and mitigating the adverse effect of heat stress on photosynthesis. Kureck et al. (2007) showed that improvements to the thermal stability of Rubisco activase by gene has shuffling increased the net photosynthesis and growth and enhanced yield in genetically transformed *Arabidopsis* under moderate heat stress. According to Schrader et al. (2004), the Rubisco activase is sensitive to temperature as low as 36 °C in cotton, resulting in to lower photosynthesis even though PS 1 cyclic electron transport increased compensating the loss of PMF for ATP formation. High-temperature-induced reduction in photorespiratory activity may reduce the adverse effect of high temperature on photosynthesis in C_3 plants.

2.4 Respiration

Temperature-mediated changes in plant respiration is an important component of the response of crop plant processes to global climate change. The Q10 is often assumed to be 2.0 (i.e., respiration doubles per 10 °C rise in temperature). Respiration, which provides the driving force for biosynthesis, cellular maintenance, and active transport in plants, involves the production of ATP and is significantly affected by the changes in temperature. This is reflected on the shifts in the control exerted by maximum enzyme activity at low temperature and substrate limitation in high temperature. According to Kokic et al. (2014), the Q10 values are lower in tissues where the respiratory flux is limited by substrate availability. The Q10 value will be lower in plants, where photosynthesis and subsequent synthesis of sugars are limited by drought and high temperature. Conversely increases in the availability of respiratory substrates due to higher photosynthesis (Under elevated CO_2) might result in an increase in the Q10 of respiration.

2.5 Biochemical Processes

Higher temperature affects the rate of biochemical reactions through the denaturation of enzymes. It is important to identify the possibility of increasing the upper limit of enzyme stability to prevent denaturation. The failure of any critical enzyme system may cause death of the plant. Most of the crop species survive high temperature up to 40–45 °C. The thermal dependence of the reaction of enzymes depends upon their Km value and stability in terms of thermal kinetic window (TKW). This TKW is determined as a result of thermal-induced lipid phase change, Rubisco activity, starch synthesis pathway in leaves, and reproductive organs. The TKW was related to dry matter production in wheat and cotton (Burkee et al. 1988). The genetic basis for heat stress tolerance in crop plants is poorly understood. The identification of adaptive quantitative trait loci (QTLs) for high-temperature tolerance is one approach being taken to try to close these knowledge gaps. For example, multiple loci for heat tolerance have been identified in wheat (grain filling) (Yang et al. 2002) and maize (pollen heat tolerance) (Frova and Sari-Gorla 1994).

Lee et al. (1983) considered heat stress may be due to the changes in the accumulation of reactive oxygen species. Peroxidation of membrane lipids has been observed at high temperature, which is a symptom of cellular injury. The adverse effect of higher-temperature stress on membranes of cell organelles and plasmalemma appears to be by denaturation of membrane proteins and melting of membrane lipids. Enhanced synthesis of an antioxidant by plant tissues may increase cell tolerance to heat (Mishra and Singhal 1992; Upadhyay et al. 1990, 1991). Exposure to high temperature brings about significant reduction in total lipid content and the ratio of unsaturated to saturated fatty acids in *Arabidopsis* (Somerville and Browse 1991). Enhancement in saturated fatty acids of cell membranes increases their melting temperature and thus confers the heat tolerance. *Arabidopsis* mutant, deficient in the activity of chloroplast fatty acid W-9 desaturase, accumulates large amount of 16:0 fatty acids, resulting in greater saturation of chloroplast lipids. This increases the optimum growth temperature and induces the heat tolerance in *Arabidopsis* (Kunst et al. 1989). However, heat tolerance in cotton does not correlate with the degree of lipid saturation (Rikin et al. 1993).

Wahid et al. (2007) reported that high-temperature injury results in the production of reactive oxygen species, such as superoxide radical (O_2 –), hydroxyl radical (OH), hydrogen peroxide (H_2O_2), and singlet oxygen ($^1O_2$). Their accumulation may disrupt cell membranes due to lipid peroxidation, destruction of proteins and nucleic acids leading to unbalanced plants homeostasis and metabolism (Mittler 2002). ROS-scavenging enzymes by enhancing cellular mechanism prevent the oxidative damage under heat stress (Larkindale and Huang 2004). The nonenzymatic antioxidants violaxanthin and zeaxanthin also contribute in photoprotection. Xanthophylls including violaxanthin and zeaxanthin partition between the light-harvesting complexes and the thylakoid membranes. This brings thermostability to the membrane and reduces its vulnerability to the lipid peroxidation under high-temperature conditions. Zeaxanthin mostly located at the periphery of the light-harvesting complex brings thermostability to the membrane by preventing the ROS-induced damage to membrane lipids (Horton 2002).

2.6 Heat Shock Proteins (HSPs)

Gradual increase in temperature increases the production of these proteins (low molecular weight proteins of 15–30 KDa, HSPs). Higher plants have at least 20 and some plant species may have as many as 40 different small heat shock proteins (Verling 1991). *Saccharomyces cerevisiae* has one small heat shock protein and *Drosophila* has four (Arrigo and Landry 1994). The small heat shock protein gene in plants comprises of a large multigene family and is nuclear encoded. Lenne and Douce (1994) reported that other classes of HSPs are localized in chloroplast, endoplasmic reticulum, cytosol, and mitochondrial locations. Plesofsky-Vig et al. (1992) demonstrated that plant's small HSPs have evolved more slowly than animal's small HSPs based on branch length and plant topology. *Zea mays* HSP 17.5 is induced by heat shock and during pollen development (Atkinson et al. 1993). Dry land crops may synthesize and accumulate the substantive level of HSPs in response to leaf temperature. These low molecular weight proteins accumulate as granular structures in the cytoplasm protecting the protein synthesis mechanisms in the cell. Plant cells respond to heat stress by rapidly accumulating the heat shock proteins inside them. This has demonstrated that the thermotolerance of crops could be modified by introducing the HSP-reporter genes (Lee et al. 1995).

2.6.1 Grain Quality

There is very little information on the impact of climate change on the food quality of crops. It seems that temperature differences of only a few degrees have more pronounced effects on quality than increased CO_2, there was a significant change in the carbohydrate composition of soybean seeds with increase in day/night temperature from 18/13 °C to 33/28 °C (day/night). The sucrose concentration increased, whereas glucose, raffinose, and fructose did not change with the rise in temperature (Wolf et al. 1982). They also demonstrated that total soluble sugars and starch decreased with increase in day/night temperature, while the proportion of soluble sugars to starch decreased. The high temperature from flowering to grain maturity caused a significant reduction in the starch accumulation period in developing wheat grains. The extremely high temperature increased the proportion of A type starch granules (10–50 um diameter) and decreased the proportion of B granules (5–10 um diameter).

2.7 Crop Growth Simulation Model

The intergovernmental panel on climate change (2007) has projected that temperature increase by the end of this century is likely to be in the range of 1.8–4.0 °C. It is also likely that tropical cyclones will become more intense, with larger peak wind speeds, and heavy precipitation will continue to become more frequent. The precipitation is very likely to increase in high altitudes, whereas it is predicted to reduce in most tropical regions. It is projected that by the end of the twenty-first century, rainfall over India will increase by 15–40 % and the mean annual temperature will increase by 3–6 °C. The warming is more pronounced in land areas, with the maximum increase over northern India. Warming is relatively greater in winter and post-monsoon seasons. Quantification of the impact of climate change parameters such as increased temperature and CO_2 on crops is being studied using well-validated simulation models. Crop growth simulation models are based on a quantitative understanding of the underlying processes and integrate the effect of soil, weather, crop, and pest management factors on growth and yield. Crop simulation models coupled with GIS prove to be strong tools for not only impact, adaptation, and vulnerability assessments but also for land use change and land use plan. Projections indicate the possibility of loss of 4–5 million tonnes in wheat production with every 1 °C increase in temperature throughout the growing period with the current level of irrigation (Aggarawal 2008). The temperatures in Indo-Gangetic plains were higher by 3–6 °C in March 2004, which is equivalent to almost 1 °C per day over the whole crop season. As a result wheat crop matured earlier by 10–20 days and wheat production dropped by more than 4 million tonnes in India (Samra and Singh 2004). Recent simulation analysis indicated that maize yield in monsoon is projected to be adversely affected due to rise in atmospheric temperature, but increased rainfall can partly offset those losses, and the spatial-temporal variations of projected changes in temperature and rainfall are likely to lead to differential impacts in different regions (Byjesh et al. 2010). According to Srivastava et al. (2010), the yield loss due to higher temperature could be counteracted by the projected increase in rainfall. However, the complete amelioration of the yield loss beyond 2 °C rise may not be attained even after the doubling of rainfall. Studies conducted on soybean using CROPGRO-soybean model have projected 50 % increased yield for a doubling of CO_2 in Central India (Lal et al. 1998). Here a 3 °C rise in temperature almost offsets the positive effects of doubling CO_2. The future climate change scenario analysis showed that mustard yield is likely to reduce in both irrigated and rain-fed conditions. However, these reductions have spatial variations in different mustard-growing areas of India (Bhoomiraj et al. 2010). Although there are uncertainties in climate change impact assessment due to inherent limitations that exist in the crop simulation models, the analysis based on simulation models provides vital and useful information on regional impact assessments, potential adaptation strategies, and vulnerability assessments. Crop modeling can predict the timing of the stages when crops are sensitive to threshold temperatures; experimental studies can provide the quantitative responses that will permit the modeling effort to progress. Such analysis is being made to support the regional policies for making agriculture sector resilient to climate change.

2.8 Adaptation and Mitigation

Several agronomic adaptation strategies, to counter the impact of higher temperature on agricultural crops, have been suggested. The agronomic strategies include both short-term adjustments and long-term adaptations. Early planting in spring increases the length of the growing season. These long-season cultivars will increase their yield potential. However, Tiroli-Cepeda and Ramos (2010) indicated that high-temperature tolerance is an intricate phenomenon involving an array of physiological and biochemical processes at the whole plant and molecular levels. Processes regulated by phytohormones, accumulation, and synthesis of osmolytes (Yousfi et al. 2010) and upregulation of the antioxidant system through genetic engineering (Wang et al. 2004) are considered important for counteracting the high-temperature stress effects on crops. Among the phytohormones, ethylene, cytokinin, and salicylic acid are found ameliorating the high-temperature effects on crop plants (Iqbal et al. 2013). The antioxidant systems such as enzymes superoxide dismutase, catalase, ascorbate peroxidase, and glutathione reductase along with low molecular weight nonenzymatic antioxidants such as a-tocopherol, ascorbate, glutathione, and carotenoids reduce the concentration of reactive oxyspecies (ROS), thus lowers the oxidative stresses in plants. Sakata et al. (2010) indicated that auxin reduction is the primary cause of high-temperature injury resulting into deleterious effect on pollen development. The application of auxin helps in counteracting these pollen injuries. Similarly Larkindale and knight (2002) in their experiment using *Arabidopsis* indicated the involvement of ethylene, ABA, and salicylic acid for the protection against high-temperature-induced oxidative damage. The maintenance of greater level of cytokinin in the kernels at elevated temperature may increase the thermotolerance in maize. Osmolytes such as proline, glycine betaine (quaternary ammonium compounds), and soluble sugars protect plants from high-temperature injuries by cellular osmotic adjustment, detoxification of ROS, protection of biological membranes, and stabilization of proteins (Wahid et al. 2007; Verbruggen and Hermans 2008). According to Wahid and Close (2007), osmolytes such as sugars, glycine betaine, and proline synthesis under high temperature may buffer cellular redox potential to reduce the effect of high-temperature stress in crop plants.

Conventional breeding for high-temperature stress tolerance was not very successful due to lack of suitable sources of genes in sexually compatible gene pools, complex nature of high-temperature traits, and lack of knowledge on genetic mechanisms of the high-temperature tolerance. Genetic studies using conventional protocols and transgenic approaches have demonstrated the dynamic nature of the high-temperature tolerance in crop plants. The thermal tolerance in different components of crop plants is being regulated by different sets of genes at different stages of growth. Thus alterations in genetic patterns of plants are more reliable in making high-temperature-tolerant varieties compared to traditional plant breeding (Bohnert et al. 2006). However, sudden changes in gene expression due to rise in temperature result in the synthesis of heat shock proteins (HSPs), which brings temperature tolerance in plants. Studies demonstrated that temperature tolerance to cold is regulated by the increase in the degree of unsaturation of fatty acids, whereas that for high-temperature tolerance in plants shows increase in the degree of saturation of fatty acids. Genetic engineering can be used to produce high-temperature-tolerant plants by reducing the degree of saturation of fatty acids in membranes by silencing the enzyme W3-fatty acid desaturase, which involves the synthesis of triple bonds in fatty acids (Murakami et al. 2000). The use of fertilizers is generally adjusted to the removal of nutrients by the crop and any losses of nutrients that may occur during or between growing seasons. The projected increases in atmospheric CO_2 concentration will increase crop growth and nitrogen uptake by the crops and thus increase the need for fertilizer applications. Changes in climate may also cause larger (or smaller) losses of nitrogen through leaching and gaseous losses.

Simple management options to counteract the warming effect are changes in sowing dates and the use of longer-season cultivars (Olesen et al. 2000; Tubiello et al. 2000). The warming effect is counteracted by the CO_2 fertilization, which will also lead to increased symbiotic nitrogen fixation in pulses (Serraj et al. 1998). The cropping area of the cooler season seed crops (e.g., pea, faba bean, and oil seed rape) and cereals (wheat and maize) will probably expand northward into Fenno-Scandinavia in Eastern Europe due to rising temperature. There will also be a northward expansion of warmer season seed crops (e.g., soybean and sunflower). An analysis of the effect of climatic change on soybean yield for selected sites in Western Europe suggests mainly an increase in yield (Wolf 2000). Yield reductions with increasing temperature have been predicted for Eastern Europe, and the yield variability may increase, especially in the steppe regions (Alexandrov 1997; Sirotenko et al. 1997).

Moving crops poleward seems an inevitable element of the multifaceted adaptation to increasing temperatures that must be implemented. However, adapting crops in the highest priority regions will require greater investment, the integration of new technologies with conventional selection-based breeding and the coordinated involvement of public and private sectors of the agricultural enterprise. Current and future increases in temperature are perhaps the most significant and most urgent challenge for the adaptation of crops to rise in temperature.

2.9 Thermal Tolerance

The hypothesis that the high-temperature induces spikelet injury was evaluated by enhancing the tolerance level in the crop models (Krishnan et al. 2007). The response of spikelet fertility to temperature depends on the average daily maximum temperature during the flowering period at which 50 % of the spikelets are fertile.

The average daily maximum temperature is 36.5 °C for Indica rice variety. To simulate an increase in the tolerance of spikelets to high temperature, the daily average maximum temperature was raised to 38.5 °C. The temperature during flowering was already higher in this season under the GCM scenarios. It was demonstrated that without adjusting the value of maximum temperature under GCMs, a large decrease in yield due to spikelet sterility was predicted. However, with the development of varieties with improved temperature tolerance of the spikelets and adaptation traits, the grain yield was increased more than the current scenarios.

There is ample need to develop temperature tolerance by exploring the following suitable strategies:

1. The induction and regulation of antioxidant defenses are necessary for obtaining substantial tolerance against temperature stress.
2. The development of genetically engineered plants by the introduction and/or overexpression of selected genes would be one of the feasible strategies.
3. In recent years a number of exogenous protectants such as proline, glycine betaine, nitric oxide, silicon, selenium, salicylic acid, and polyamines are identified as protectants for plants against damage from extreme temperatures.

2.10 Future Thrust

The impact assessment of many crops is still not done because of the nonexistence of crop models for these crops. The following future strategies will help in planning the productive research on crop productivity under changing temperature:

1. Developing indigenous crop simulation models such as InfoCrop is desirable. Hence there is an urgent need for developing a crop simulation model for other important crops.
2. The crop coefficients/response to temperature, CO_2, and other climatic factors needs to be better understood under stress conditions so that a robust model could be made.
3. The response of crops to sudden changes in temperature or extreme rainfall or other climatic stresses at different phenological phases needs to be quantified for reliable simulation of the impact of climatic variability on crop production.
4. Development of decision support system (DSS) for managing crops in climate change scenarios with sustainable resource management to reduce the climatic risks on farm income.

According to IPCC (2007) assessment, agricultural production in South Asia may fall about 30 % by 2050 if no action is taken to counter the effects of rising temperature. It has been suggested various options to mitigate the impacts of increasing temperature:

1. Developing varieties tolerant to heat stress
2. Modifying the crop management practices

3. Improving water management for agricultural crops
4. Diversifications of crops
5. Adopting the resource conserving new farm technologies
6. Improving pest management
7. Better weather forecasts and crop insurance
8. Harnessing the indigenous technical knowledge of farmers

2.11 Crop/Weed Interactions

Weeds reduce the crop productivity, primarily by competing for resources and sometimes by chemical interference. Changes in temperature, precipitation, and rising atmospheric CO_2 have potentially important consequences for crop/weed interactions, which is evidenced from the basic biology of crops and weeds. Based on the differences in temperature optima for physiological processes, it is predicted that C_4 species will be able to tolerate high temperature than C_3 species. Therefore, C_4 weeds may benefit more than the C_3 crops from any temperature increases that accompany elevated CO_2 levels. High CO_2 levels ameliorate the effects of suboptimal temperatures (Sionit et al. 1987) and other forms of stress on plant growth (Bazzaz 1990). O'Donnell and Adkins (2001) reported that wild oat plants grown at high temperature 23/19 °C (day/night) completed their development faster than those grown at normal temperature 20/16 °C. If the maturation is faster relative to the crop, more seeds may be deposited in the soil seed bank with a consequent increase in the number of wild oat plants. The reduction of growth caused by water stress and nutrient deficiency may reduce the impact of weeds on crop production, since smaller plants interfere less with each other (Patterson 1995). The increase in temperature allows some of the sleeper weeds to become invasive. Huge environmental damage and control cost can be prevented if these weeds are eradicated before they become widespread. Species in cool locations with restricted range can persist by reproducing only in warmer places, but increasing temperatures might allow this weed to increase seed production rapidly, become invasive, and expand its range into more favorable habitats. In other cases, weed problem might diminish in some areas or shift as a result of change in temperature. The physiological plasticity of weeds and their greater intraspecific genetic variation compared with most crops would provide weeds with a competitive advantage in an elevating temperature condition. Weeds with greater genetic diversity may become problematic especially under higher temperature and rising atmospheric CO_2. The fact that many weeds are C_4 and many crops are C_3 may seem advantageous, but will be of little comfort to those trying to grow C_4 crops in competition with weeds. It was also observed that due to increasing temperature and rising CO_2 concentration, some of the tropical and subtropical weeds extend their range toward the poles and become troublesome in areas where they are not currently a problem. There may be a shift in the existing weed flora, which necessitates changes in the planning and implementation of weed control programs. The changes in temperature along with the higher concentration of CO_2 also have been found to alter the herbicide efficacy.

High temperatures limit reproductive development and decrease reproductive output more likely in C_3 species. This could alter the weed community composition and affect crop/weed interactions. The mean temperatures in temperate regions are normally below optimum temperature for vegetative growth even for C_3 species. To reduce the impact of weed populations on crop growth and yield in temperate regions, planting times must be adjusted such that crops and weeds would be exposed to the same temperatures during the establishment phase as they are currently growing even if there is a rise in temperature. Higher temperatures would increase the evaporative demand. C_4 species, with their high water use efficiency and CO_2-saturated photosynthesis, are better adapted to high evaporative demand. This will markedly increase the crop/weed competition. Though the crop scientists have tried to eliminate weeds as a variable in their experiments, the field studies need to understand the interactive responses of weeds as well as crops to the climate changes. Such information is necessary to predict how changes in climate may alter weed populations and crop losses due to weeds (Bunce and Ziska 2000).

Weeds have tremendous impact on agricultural productivity in the tropics. It is estimated that weeds contribute to loss in crop yields to the extent of 25 %. Yield losses in cereal crops ranged from 0.4 to 15 % in marginally weeded crops, whereas the loss ranges from 1 to 32 % in chemical weed control and to the extent of 72.5 % in unweeded conditions compared to completely weed-free condition. It is of interest to know that negative effect is more in kharif crops, many of which are C_4. Integrated weed management technologies, which are safe to the crop and environment, required to be based on climatic factors at various agroecozones assume great importance. Further, the possibility of using bio-herbicides and high-specificity herbicides is the new nontoxic remedial measures to counter the weed crop competition at the critical stages of vulnerability to higher temperatures.

2.12 Biotic Stresses and Pest Management

Global climate changes may affect the crop yields, incidence of pests, and economic cost of crop production. The probable effects of rise in temperature on pests may include expansion of their ranges, change in population growth rates, increased period of activity, alteration in crop-pest synchrony and natural enemy-pest interaction, and changes in interaction among pest species (Southerst et al. 1996; Hughes 2000; Walther et al. 2002; Root et al. 2003). The effects of temperature increases on insects are likely to be more pronounced in the Arctic and other extreme environments and least in the tropics. The individuals may develop faster at high temperature and survival may even be enhanced, but these insects may consequently have lower adult weight and fecundity (Bale et al. 2002). Global warming will lead to a higher incidence of pests and diseases and thus to a potentially larger use of pesticides. The use of pesticides, however, needs to be kept low through the adoption of integrated pest management systems, which targets the control measures to the observed problem.

The insect pest buildup during the growing season of the crops depends upon their survival during extreme temperatures, and their population increase during the crop season. High temperature during crop's growing season aggravates the development rates of insects and pests. Plant breeders can contribute by developing cultivars with resistance to various insects and pests. Changes in community structure and extinction of some species are also expected (Thomas et al. 2004). Pest management involves intensive decision-making and effective decision support tools. Pest management research is required to derive practical tools to develop tactics and strategies for pest management. System approach provides such tools in the form of simulation and decision models. Simulation models help to derive such tools and decision models extend support to such tools. Crop growth simulation models take into consideration the physiological basis of pest damage for simulating crop-pest interactions (Pinnschmidt et al. 1995).

Temperature is the major factor in global climate change that directly affects insect development, reproduction, and survival (Parmesan 2007; Van Asch and Visser 2007). The fossil record of leaves that fell off trees about 55 Ma ago during the Palaeocene-Eocene Thermal Maximum (PETM) period indicated 5 °C rise in global temperature caused by the tripling of CO_2 level. During this period the insect numbers were predicted to be soaring and left an indelible impression on the fossilized leaves preserved since that time. It was found that as the temperature increased, the leaves looked more nibbled (Hopkin 2008). The effect of higher temperature in general has been predicted to an increase intensity of herbivore pressure on plants. With every degree rise in global temperature, the life cycle of the insect will be shorter. The quicker the life cycle, the higher will be the population of pests. Bale et al. (2002) predicted that if the global temperature increases, the species will shift their geographical ranges closer to the poles or to higher elevations and increase their population size. They have concluded that the rise in temperature would affect crop-pest insect populations by:

1. Extension of geographical range
2. Increased overwintering
3. Changes in population growth rate
4. Increased number of generations
5. Extension of development seasons
6. Change in crop-pest synchrony
7. Changes in interspecific interactions
8. Increase in the risk of invasion by migrant pests
9. Introduction of the overwintering hosts and other alternatives.

Temperature can exert different effects depending on the developmental strategy of the insect species (Bale et al. 2002). Temperature changes can affect insect physiology and development directly or indirectly. Some insects take several years to complete one life cycle. These insects (cicadas, arctic moths) will tend to moderate temperature effects over the course of their life history. Some crop pests are "stop and go" developers in relation to temperature, so they develop more rapidly during periods of suitable temperatures. Migratory insects may arrive earlier or the area in which they are able to overwinter may be expanded. Parasitism could be reduced if host populations emerge and pass through vulnerable life stages before parasitoids emerge. Hosts may pass through vulnerable life stages more quickly at higher temperatures, reducing the window of opportunity for parasitism. The temperature may change gender ratios of some pest species such as thrips (Lewis 1997) potentially affecting the reproduction rates. Insects, which spend the larger part of their life in the soil, may be more slowly affected by temperature changes than those that are aboveground because soil provides an insulating medium which will buffer the temperature changes, more than the air. Mild winters and longer summers are highly favorable to increased aphid populations due to rise in mean temperature, resulting in to an extension of the geographical range. There is also the prospect of new pests, which may become much more important as a result of increased temperature. The increasing winter temperatures have been considered to be the key factor for the shifting of insects for reducing winter mortality (Battisi et al. 2005). According to a study conducted on 1100 insect species, rising temperature may cause 15–37 % of those species to extinct by 2050 (Thomas et al. 2004; Hance et al. 2007). In another survey of 1600 species, about 940 showed the effects of global warming. For example, in Europe, 35 species of butterflies have already shifted their ranges 35–240 km northward. 70 % of 23 butterfly species now start their first flight about 24 days earlier than they used to do 31 years ago in California (Parmesan and Yohe 2003; Parmesan 2007). Parmesan and Yohe (2003) reported that more than 1700 northern hemisphere species have exhibited shifts averaging 6.1 km per decade toward the poles. At the same time, effects of defoliators, wood borers, and bark beetles could become more detrimental due to prolonged growing season leading to multivoltinism (Stastny et al. 2006). Parmesan and Yohe (2003) estimated that about 59 % of 1598 species investigated exhibited measurable changes in their phenologies and/or distribution over the past 20–140 years. They estimated advancement of spring events by 2.3 days/decade based on their phenological responses. Root et al. (2003) estimated an advancement of 5.1 days/decade in a similar study. Degree day or phenology-based models are often used to predict the emergence of insects like cabbage maggot, onion maggot, European corn borer, Colorado potato beetles, etc. and their potential to damage crops.

Changes in rainfall pattern also affect insect survival. Small insects like aphids on wheat crop have been adversely affected by rainfall; however, more intense rainfall may reduce incidence of small pests in crops (Chander 1998). It is predicted that the 1 °C increase in temperature would enable species to spread 200 km northward or 140 km upward in altitude (Parry et al. 1989). Climate warming may result in a majority of temperate insects to extend their ranges to higher latitudes and altitudes. According to Hill (1987), the minimum temperature plays a greater role in the distribution of insect species. However, the global warming may affect migration and spreading of the pests toward further north. The rise in the temperature affects delay in the onset of hibernation thus increasing the period of activity of pests. Plants of Rape seed mustard are infested by two aphid species *Lipaphis erysimi* and *Myzus persicae*, the former being dominant during severe winters whereas the latter during mild winters. With rise in temperature, higher incidence of *Myzus persicae* is generally witnessed (Chander and Phadke 1994). Pollard and Yates (1993) reported that in multivoltine species such as Aphididae and some Lepidoptera, higher temperatures allow faster development rate probably for additional generations within a year.

The increase in temperature significantly alters host physiology and resistance. High temperature may cause breakdown temperature-sensitive resistance. The plant defensive system is lowered and they become vulnerable to pest attack. However, in some forage species, there is increased lignification at a higher temperature that enhances the resistance to pathogens. Rising temperature will affect the geographical distribution of plant species and their growth pattern and would bring spatial changes in crops and cropping systems for sustained productivity. New pests may arise while some pests may become less important if global warming results in the northward shift of agroclimatic zones and host plants migrate into new regions. The capacity of an herbivore insect to complete its development depends on the adaptation to both the environmental conditions and the host plants. The changed temperature, which promotes the expansion of insect's range, may also involve a new association between an herbivore and its host. The large outbreaks observed in the expansion areas on the new hosts may be explained either by the high susceptibility of the hosts or by the inability of natural enemies to locate the moth larvae on an unusual hosts or environment (Stastny et al. 2006).

Sutherst et al. (1996) and Teng and Yang (1993) have suggested model-based assessment of the effects of climate change on insects and pest population. Variables such as monthly minimum and maximum temperatures, precipitation, and evaporation are being used for impact assessment of the biotic stresses on crop plants. Based on these variables, Bag et al. (1991) predicted that climate warming could result in increased nematode and virus problems in northern Europe. However, Chander et al. (2003) observed that the rise in temperature and more intense rains in future might reduce aphid incidence in barley in India. The model developed for rice gundhi bug (*Leptocorisa acuta*) demonstrated that up to 1 °C rise in daily average temperature over the present temperature of Delhi would not affect gundhi bug population much, but a further increase would cause an appreciable decline in it.

Pest and disease problem is more severe in tropical regions due to global warming. The incidence of pest and diseases deleteriously affects the crop yield. Climatic factors are the causative agents in determining the population fluctuations. Considering the breeding seasons for desert locust, it was seen that where rainfall occurred during winter and early spring (Southeast Arabia, Southern Iran, and Baluchistan), the locust breeds in spring. The swarms produced in winter and spring in the Middle East and Baluchistan usually fly eastward in to Pakistan and India during summer season. With the advent of monsoon, they lay eggs from June to September. They invade all parts of India and damage kharif crops (Rao and Rao 1996). The possibility of forecasting of the spread of aphids (*Lipaphis erysimi*) on the mustard crop grown during winter in northern India due to western disturbances was established. These western disturbances cause cold and humid air resulting in cloudy weather favorable for the occurrence of aphids on the mustard crop.

The efficiency of pest control operations depends on weather factors. Dusting and spraying of pesticides require 18–24 h of rain-free period after chemical treatment. Considerable use of meteorological data in forecasting the occurrence of pests on food crops for timely pest control and minimum use of pesticides is established. The role of microclimate at ground level and higher elevation in long-distance dispersal of pathogens leading to epidemic needs to be analyzed to develop preventive measures.

Further stress needs to be given to following parameters for integrated insect pest management for global warming conditions:

1. Impact on insect feeding behavior based on host plant metabolism and chemical composition such as C: N ratio and accumulation of secondary metabolite like phenols, associated with the allelopathy among plants and its impact on resistance to insects and pathogens.
2. Modification in the susceptibility of pathogen attack due to climate change-induced biochemical and structural changes in the host.
3. High-temperature stress in host plants could increase the vulnerability to pest attack.
4. Alteration in the geographical distribution of agricultural insects and pests upsets the natural control of predator and diseases and increased invasion by migratory exotic pests (Idso et al. 1987).

2.13 Adaptation to High-Temperature Stress

Some adaptation measures for combating pest menace under climate change are:

1. Alteration in sowing dates of crops may alter host-pest synchrony. There is a need to explore changes in pest-host interaction under agronomic management adaptation.
2. Breakdown of temperature-sensitive resistance may lead to a more rapid evolution of pest biotypes. This would require breeding of cultivars for both pest and drought resistance.
3. Temperature-tolerant strains of natural enemies of pests should be evolved to take the advantage of the rise in atmospheric temperature.
4. Global warming could affect the efficacy of crop protection chemicals through:
   (a) Changes in temperature and rainfall.
   (b) Morphophysiological changes in crop plants (Coakley et al. 1999). Intense rainfall could increase pesticide wash off and reduced pest control. The increased thickness of epicuticular wax layer under high CO_2 reduces the uptake of pesticide by the host, while the increased canopy size causes dilution of active ingredients in host tissues. On the contrary increased metabolic rate at a higher temperature could result in faster uptake of pesticides by plants and higher toxicity to pests. Thus the rates of pesticide application have to be modified according to the situations.

5. Pest forecasting using reliable medium-range weather forecast can help in cautioning farmers about impending the pest situation and adoption of preventive measures to avert pest problems. The determination of economic injury levels (EILs) for pests is based on yield infestation relationships that are site specific with little scope for extrapolation. Pest dynamics-crop models will be useful tools to assess an impact of climate change on pest dynamics and crop productivity. Elevated temperature will alter the ranges and abundances of insects and, therefore, have profound impacts on agriculture by the movement of existing crop pests into new areas by rising currently disregarded insect species to pest status.

References

Aggarawal PK (2008) Global climate change and Indian agriculture: impacts, adaptation and mitigation. Indian J Agric Sci 78:911–919

Al Khatib K, Poulson GM (1999) High temperature effects on photosynthetic processes in temperate and tropical cereals. Crop Sci 39:119–125

Alexandrov VA (1997) Vulnerability of agronomic systems in Bulgaria. Climate Change 36:135–149

Allen LH Jr, Vu JCV, Valle RR et al (1988) Nonstructural carbohydrates and nitrogen of soybean grown under carbon dioxide enrichment. Crop Sci 28:84–94

Arrigo AP, Landry J (1994) 14 expression and function of the low-molecular-weight heat shock proteins. Cold Spring Harb Monogr Arch 26:335–373

Atkinson BG, Raizada M, Bouchard RA et al (1993) The independent stage specific expression of the 18-kDA heat shock protein genes during microsporogenesis in *Zea mays* L. Dev Genet 14:15–26

Bagga AK, Rawson HM (1977) Contrasting responses of morphologically similar wheat cultivars to temperatures appropriate to warm temperate climates with hot summers: a study in controlled environment. Aust J Plant Physiol 4:877–887

Baker JT, Allen LH Jr, Boole KJ et al (1989) Response of soybean to air temperature and carbon dioxide concentration. Crop Sci 29:98–105

Bale JS, Masters GJ, Hodkinson ID et al (2002) Herbivory in global climate research: direct effect on rising temperature on insect herbivores. Glob Change Biol 8:1–16

Battisti A, Stastny M, Netherer S et al (2005) Expansion of geographic range in the pine processionary moth caused by increased winter temperatures. Ecol Appl 15(5):2084–2096

Bazzaz FA (1990) The response of natural ecosystems to the rising global CO_2 levels. Annu Rev Ecol Syst 21:167–196

Bhattacharya S, Bhattacharya NC, Biswas PK et al (1985) Response of cow pea (Vigna unguiculata L.) to CO2 enrichment environment on growth, dry matter production and yield components at different stages of vegetative and reproductive growth. J Agric Sci 105:527–534

Bhoomi Raj K, Chakrabarti B, Aggarwal PK et al (2010) Assessing the vulnerability of Indian mustard to climate change. Agric Ecosyst Environ 138:265–273

Boag B, Crawford JW, Neilson R (1991) The effect of potential climatic changes on the geographical distribution of the plant-parasitic nematodes xiphinema and longidorus in Europe. Nematologica 37(1):312–323

Bohnert HJ, Gong Q, Li P (2006) Unraveling abiotic stress tolerance mechanisms: getting genomics going. Curr Opin Plant Biol 9:180–188

Bunce JA, Ziska LH (2000) Crop ecosystem responses to climate change. In: Reddy KR, Hidges HF (eds) Crop/weed interactions: climate change and global crop productivity, CAB International, pp 333–352

Burkee JJ, Mahan JR, Hatfield JL (1988) Crop specific thermal kinetic windows in relation to wheat and cotton biomass production. Agron J 80:553–556

Byjesh K, Naresh Kumar S, Aggrawal PK (2010) Simulating impacts, potential adaptation and vulnerability of maize to climate change in India. Mitig Adapt Strat Glob Chang 15:413–431

Camejo D, Rodiguez P, Morales MA et al (2005) High temperature effects on photosynthetic activity of two tomato cultivars with different heat susceptibility. J Plant Physiol 162:281–289

Carney KM, Hungate BA, Drake BG et al (2007) Altered soil microbial community at elevated CO_2 leads to loss of soil carbon. Proc Natl Acad Sci 104:4990–4995

Chander S (1998) Infestation of root and foliage/earhead aphids on wheat in relation to predators. Indian J Agric Sci 68(11):754–755

Chander S, Phadke KG (1994) Incidence of mustard aphid, Lipaphis erysimi and potato aphid, Myzus persicae on rapeseed crop. Ann Agric Res 15(3):385–387

Chander S, Singh VS, Kalra N (2003) Aphid infestation on barley in relation to climate variability. In: Proceedings of the national symposium on frontier areas of entomological research. IARI, New Delhi, pp 37–38

Craufurd RQ, Wheeler TR (2009) Climate change and flowering time of annual crops. J Expt Bot 60:2529–2539

Coakley SM, Scherm H, Chakraborty S (1999) Climate change and plant disease management. Ann Rev Phytopathol 37:399–426

De Ronde JAD, Cress WA, Kruger GHJ et al (2004) Photosynthetic response of transgenic soybean plants containing an Arabidopsis P5CR gene, during heat and drought stress. J Plant Physiol 61:1211–1244

Egeh AO, Ingram KT, Zamora OB (1994) High temperature effects on leaf exchange. Philipp J Crop Sci 17:21–26

Frova C, Sari Gorla M (1994) Quantitative trait loci (QTLs) for pollen thermotolerance detected in maize. Mol Genet Genomics 245:424–450

Gifford RM (1977) Growth pattern, CO2 exchance and dry weight distribution in wheat growing under different photosynthetic environments. Aust J Plant Physiol 4:99–110

Gifford RM (1989) Trees in the global greenhouse. Their role and potential. Trees Nat Resour 31:9–11

Hall AE (1992) Breeding for heat tolerance. Plant Breed Rev 10:129–168

Hance T, Baaren J, Vanvernon P et al (2007) Impact of extreme temperatures on parasitoides in a climate change perspective. Ann Rev Entomol 52:107–126

Hedhly A, Hormaza JI, Herrero M (2008) Global warming and sexual plant reproduction. Trends Plant Sci 14:30–36

Hesketh JD, Hellmers H (1973) Floral initiation in four plant species growing in CO2 enriched air. Environ Control Biol 11:51–53

Hill DS (1987) Agricultural insects, pests of temperate regions and their control. Cambridge University Press, Cambridge, p 659

Hopkin K (2008) Plants don't like greenhouse effect. http://www.scientificamerican.com/podcast/episode/12c9008d-fb71-7c7b-ebadb04daeb4356c/. Accessed 25 Feb 2016

Horton P (2002) Crop improvement through alteration in the photosynthetic membrane, ISB news report. Virginia Tech, Blacksburg

Hughes L (2000) Biological consequences of global warming: is the signal already apparent? Trends Ecol Evol 15:56–61

Idso SB, Kimball BA, Mauni JR (1987) Atmospheric carbon dioxide enrichment effects on cotton mid day foliage temperature: implications for plant water use and crop yield. Agron J 79:667–672

Imai K, Coleman DF, Yanagisawa T (1985) Increase in atmospheric partial prssure of carbon dioxide on growth and yield of rice (Oryza sativa). Jpn J Crop Sci 53:413–418

IPCC (International Panel on Climate Change) (2007) Summary for policy makers. In: Parry ML et al (eds) Climate change 2007: impacts, adaptation and vulnerability, contribution of working group II to the fourth assessment report of the Intergovernmental Panel on Climate Change. Cambridge University Press, Cambridge

Iqbal M, Khan R, Asgher M et al (2013) Rising temperature in the changing environment: a serious threat to plants. Clim Chang Environ Sustain 1:25–36

June T, Evans JR, Farquhar GD (2004) A simple new equation for the reversible temperature dependence of photosynthetic electron transport: a study on soybean leaf. Funct Plant Biol 31:275–283

Kendall AC, Turner JC, Thomas SM (1985) Effects of CO2 enrichment at different irradiances on growth and yield of wheat. I, Effects of cultivar and duration of CO2 enrichment. J Exp Bot 36:252–260

Kokic P, Crimp S, Howdon M (2014) A probabilistic analysis of human influence on recent record global mean temperature changes. Clim Risk Manage 3:1–12

Krenzer EG, Moss DN (1975) Carbon dioxide enrichment effects upon yield and yield components in wheat. Crop Sci 15:71–74

Krishnan P, Swain DK, Basker C et al (2007) Impact of elevated CO_2 and temperature on rice yield and methods of adaptation as evaluated by crop simulation studies. Agric Ecosyst Environ 122(2):233–242

Kunst L, Browse J, Somerville C (1989) Enhanced thermal tolerance in a mutant of Arabidopsis deficient in palmitic acid unsaturation. Plant Physiol 91:401–408

Kureck I, Chang TK, Bertain SM et al (2007) Enhanced thermostability of arabidopsis rubisco activase improves photosynthesis and growth rates under moderate heat stress. Plant Cell 19:3230–3241

Lal M, Singh KK, Rathore LS et al (1998) Vulnerability of rice and wheat yields in NW India to future changes in climate. Agric Forest Meteorol 89:101–114

Larkindale J, Huang B (2004) Thermotolerance and anti oxidant systems in agrostis stolonifera. Involvement of salicylic acid, abscisic acid, calcium, hydrogen peroxide and ethylene. J Plant Physiol 161:405–413

Larkindale J, Knight MR (2002) Protection against heat stress induced oxidative damage in Arabidopsis involves calcium, abscisic acid, ethylene and salicylic acid. Plant Physiol 128:682–695

Lecain D, Smith D, Morgan J et al (2015) Microclimatic performance of a free air warming and CO_2 enrichment experiment in Windy Wyoming, USA. PLOS one 6:1–14. doi:10.1371/journal.pone.0116834

Lee PC, Bochner BR, Ames BN (1983) A heat shock stress and cell oxidation. Proc Natl Acad Sci 80:7496–7500

Lee GJ, Pokala N, Vierling E (1995) Structure and *in vitro* molecular chaperone activity of cytosolic small heat shock proteins from pea. J Biol Chem 270:10432–10438

Lenne C, Douce R (1994) A low molecular mass heat-shock protein is localized to higher plant mitochondria. Plant Physiol 105:1255–1261

Lewis T (1997) Thrips as crop pests. CAB International, Oxon

Luo C, Xu G, Chao Z et al (2010) Effect of warming and grazing on litter mass loss and temperature sensitivity of litter and dung mass loss on the Tibetan plateau. Glob Change Biol 16:1606–1617

Maheswari M, Joshi DK, Saha R et al (1999) Transverse relaxation time of leaf water protons and membrane injury in wheat (*Triticum aestivum* L.) in response to high temperature. Ann Bot 84:741–745

Mishra RK, Singhal GS (1992) Function of photosynthetic apparatus of intact wheat leaves under high light and heat stress and its relationship with thylakoid lipids. Plant Physiol 98:1–6

Mittler R (2002) Oxidative stress, antioxidants and stress tolerance. Trends Plant Sci 7:405–410

Murakami Y, Tsuyama M, Kobayashi Y (2000) Trienoic fatty acids and plant tolerance for high temperature. Science 287:476–479

O'Donnell CC, Adkins SW (2001) Wild oat and climate change: the effect of CO_2 concentration, temperature, and water deficit on the growth and development of wild oat in monoculture. Weed Sci 49(5):694–702

Olesen JE, Jensen T, Petersen J (2000) Sensitivity of field-scale winter wheat production in Denmark to climate variability and climate change. Clim Res 15:221–238

Parmesan C (2007) Influences of species, latitudes and methodologies on estimates of phonological response to global warming. Global Change Biol 13:1860–1872

Parmesan C, Yohe G (2003) A globally coherent fingerprint of climate change impacts across natural systems. Nature 421:37–42

Parry DA, Amaranthus MP, Borchers JG et al (1989) Boot strapping in ecosystems. Bioscience 39:230–237

Patterson DT (1995) Weeds in a changing climate. Weed Sci 43:685–701

Pinnschmidt HO, Batchelor WD, Teng PS (1995) Simulation of multiple species pest damage on rice. Agric Syst 48:193–222

Plesofsky-Vig N, Paulson A, Hill EP et al (1992) Heat shock gene expression in germinating ascospores of *Neurospora tetrasperma*. FEMS Microbiol Lett 90:117–122

Pollard E, Yates TJ (1993) Monitoring butterflies for ecology and conservation. Chapman and Hall, London

Porter JR, Semenov MA (2005) Crop responses to climate variation. Philos Trans Royal Soc 360:2021–2035

Prasad PVV, Boote KJ, Allen LH Jr (2002) Effects of elevated temperature and carbon dioxide on seed set and yield of kidney bean (*Phaseolus vulgaris* L.). Glob Change Biol 8:710–721

Prasad PV, Boote KJ, Allen LH Jr et al (2003) Super optimal temperatures are detrimental to peanut (*Arachis hypogea*) reproductive processes and seed yield at both ambient and elevated carbon dioxide. Glob Change Biol 9:1775–1787

Rao BVR, Rao MR (1996) Weather effects on pests. In: Abrol YP, Gadgil S, Pant GB (eds) Climate variability and agriculture. Narosa Publication House, New Delhi, pp 281–296

Reddy KR, Hodges HF, Kimball BA (2000) Crop ecosystem response to climate change: cotton In: Reddy KR (eds) Climate change and global crop productivity. CAB International, pp 161–187

Rekika D, Nachit MM, Monneveux P (1997) Comparative osmotic adjustments in barley and tetraploid wheats. Plant Breed 116(6):519–523

Rikin A, Dillworth JW, Bergman DK (1993) Correlation between circadian rhythm of resistance to extreme temperature and changes in fatty acid composition in cotton seedlings. Plant Physiol 101:31–36

Root TL, Price JT, Hall KR et al (2003) Finger prints of global warming on wild animals and plants. Nature 421:57–60

Rustad LE, Campbell JL, Marion GM et al (2001) A meta analysis of the response of soil respiration, net nitrogen mineralization and above ground plant growth to experimental ecosystem warming. Oecologia 126:543–562

Sage RF, Kubian DS (2007) The temperature response of C3 and C4 photosynthesis. Plant Cell Environ 30:1086–1106

Saini HS, Aspinall D (1982) Abnormal sporogenesis in wheat (*Triticum aestivum* L.) induced by short period of high temperature. Ann Bot 49:835–846

Sakata T, Oshino T, Miura S (2010) Auxins reverse plant male sterility caused by high temperatures. Proc Natl Acad Sci 107(19):8569–8574

Salvucci ME (2008) Association of Rubisco activase with chaperonin-60beta; a possible mechanism for protecting photosynthesis during heat stress. J Exp Bot 59:1923–1933

Samara JS, Singh G (2004) Heat wave of March, 2004: impact on agriculture, Indian council of agricultural research, p 32

Santarius KA (1973) The protective effect of sugars on chloroplast membranes during temperature and water stress and its relationship to frost, desiccation and heat resistance. Planta 113:105–114

Satake T, Yoshida S (1978) High temperature induced sterility in Indica rice at flowering. Jpn J Crop Sci 47:6–10

Schrader SM, Wise RR, Wacholtz WF (2004) Thylakoid membrane responses to moderately high leaf temperature in Prima cotton. Plant Cell Environ 27(6):725–735

Serraj R, Sinclair TR, Allen LH (1998) Soybean nodulation and N_2 fixation response to drought under carbon dioxide enrichment. Plant Cell Environ 21:491–500

Sionit N, Strain BR, Flint EP (1987) Interactions of temperature and CO_2 enrichment on soybean: Growth and dry matter partitioning. Can J Plant Sci 67:59–67

Sirotenko OD, Abashina HV, Pavlova VN (1997) Sensitivity of the Russian agriculture to changes in climate, CO_2 and tropospheric ozone concentrations and soil fertility. Climate Change 36:217–232

Somerville C, Browse J (1991) Plant lipids, metabolism and membranes. Science 252:80–87

Southerst RW, Yonow T, Chakraborty S (1996) A generic approach to defining impacts of climate change on pests, weeds and diseases in Australasia. In: Bouma WJ (ed) Greenhouse coping with climate change. CSIRO, Collingwood, pp 281–307

Srivastava A, Naresh kumar S, Aggarwal PK (2010) Assessment on vulnerability of sorghum to climate change in India. Agric Ecosyst Environ 138:160–169

Stastny M, Battisti A, Petrucco Toffolo E (2006) Host plant use in the range expansion of the pine processionary moth Thaumetopoea pityocampa. Ecol Entomol 31:481–490

Sujatha KB, Uprety DC, Nageswara Rao D (2008) Up regulation of photosynthesis and sucrose-P synthase in rice under elevated carbon dioxide and temperature conditions. Plant Soil Environ 54:155–162

Sutherst RW, Yonow T, Chakraborty S et al (1996) A generic approach to defining impacts of climate change on pests, weeds and diseases in Australasia. In: Bouma WJ et al (eds) Greenhouse, coping with climate change. CSIRO, Melbourne, pp 281–307

Teng PS, Yang XB (1993) Biological impact and risk assessment in plant pathology. Ann Rev Phytopathol 31:495–521

Thomas JA, Telfer MG, Roy DB et al (2004) Comparative losses of British butterflies, birds and plants and the global extinction crisis. Science 303:1879–1881

Tiroli-Cepeda AO, Ramos CH (2010) Heat causes oligomeric disassembly and increases the chaperone activity of small heat shock proteins from sugarcane. Plant Physiol Biochem 48(2–3):108–116

Tubiello FN, Donatelli M, Rosenzweig C et al (2000) Effects of climate change and elevated CO2 on cropping systems: model predictions at two Italian locations. Eur J Agron 13:179–189

Upadhyay A, Davis TD, Larson MH et al (1990) Uniconazole induced thermotolerance in soybean seedling root tissue. Physiol Planta 79:78–84

Upadhyay A, Davis TD, Sankhla M (1991) Heat shock tolerance and anti oxidant activity in moth bean seedlings treated with tetayclasis. Plant Growth Regul 10:215–222

Uprety DC, Dwivedi N, Raj A et al (2009) Study on the response of diploid, tetraploid and hexaploid species of wheat to the elevated CO2. Physiol Mol Biol Plants 15(2):161–168

Uprety DC, Garg SC, Tiwari MK et al (2000) Crop responses to elevated CO2Technology and Research (Indian studies). Global Environ Res 3:155–167

Uprety DC, Mishra RS, Abrol YP (1995) Effect of elevated CO2 and moisture stress on the photosynthesis and water relation in Brassica species. J Crop Sci Agron 175:231–237

Van Asch M, Visser ME (2007) Phenology of forest caterpillars and their host trees: the importance of synchrony. Annu Rev Entomol 52:37–55

Verbruggen N, Hermans C (2008) Proline accumulation in plants: a review. Amino Acids 35:753–759

Vierling E (1991) The roles of heat shock proteins in plants. Annu Rev Plant Physiol Plant Mol Biol 42:579–620

Wahid A, Close TJ (2007) Expression of dehydrin under heat stress and their relationship with water relations of sugar cane leaves. Biol Planta 51:104–109

Wahid A, Gilani S, Ashraf M (2007) Heat tolerance in plants: an overview. Environ Expt Bot 61:199–223

Wall GW, Mclain ET, Kimball BA (2013) Infrared warming effects intra row soil CO_2 efflux during vegetative growth of spring wheat. Agron J 105:607–618

Walther GR, Post E, Convey P (2002) Ecological responses to recent climate change. Nature 416(6879):389–395

Wise RR, Olson AJ, Schrader SM et al (2004) Electron transport is the functional limitation of photosynthesis in field grown Pima cotton plants at high temperature. Plant Cell Environ 27:717–724

Wolf RB, Cavins JF, Kleiman R (1982) Effect of temperature on soybean seed constituents: oil, protein, moisture, fatty acids, amino acids and sugars. J Am Oil Chem Soc 59:230–232

Wolf J (2000) Modelling climate change impacts at the site scale on soybean. In: Downing TE, Harrison PA, Butterfield RE et al (eds) Climate change, climatic variability and agriculture in Europe, environ-mental change unit. University of Oxford, Oxford, pp 103–116

Xu Q, Paulson AQ, Guikema JA et al (1995) Functional and ultrastructural injury to the photosynthesis in wheat by high temperature during maturation. Environ Expt Bot 35:43–54

Yang J, Sears RG, Gill BS et al (2002) Quantitative and molecular characterization of heat tolerance in hexaploid wheat. Euphytica 126:75–282

Yousfi N, Slama I, Ghnaya T et al (2010) Effects of water deficit stress on growth, water relations and osmolyte accumulation in Medicago truncatula and M. laciniata populations. C R Biol 333(3):205–213

Wang W, Vinocur B, Shoseyov O et al (2004) Role of plant heat-shock proteins and molecular chaperones in the abiotic stress response. Trends Plant Sci 9(5):244–252

Temperature Enrichment Technologies for Crop Response Studies

3

Various types of facilities are developed to study the impact of rising surface temperature on the responses of crops.

3.1 Temperature Gradient Chambers

These temperature gradient chambers (TGCs) are constructed to study the responses of crop plants to the interaction of the rise in temperature and atmospheric CO_2. Temperature gradient chambers are 50 cm high and 10 m long plastic greenhouse with an air inlet at one end and an exhaust fan at the other end. These are used as commercial greenhouses with steel pipes of 19 mm in diameter. Dimensions of the chamber are also of 3.6 m high × 18 m length. These are covered with 0.1 mm thick UV transparent polyvinyl chloride film. TGCs are kept in north-south orientation to obtain a fairly uniform spatial distribution of direct solar beam in summer and to minimize the influence of shading of neighboring chambers in winter. Light transmittance observed under the clear sky is 77 % in July and 66 % in November. Perforated panels are used as air inlets to regulate the incoming air flow. Three exhaust fans are installed at other ends each with ventilation capacity of 28 m^3min^1. The mean air flow velocity in transverse section is 26, 17, and 9 cm S^1 in this sequence, respectively, for these fans (Horie et al. 1995). Horie et al. (1991) developed such a greenhouse tunnel which is about 30 m long, 3 m wide, and 2 m in height. Natural solar energy is used to create temperature gradients in the greenhouse tunnel. The air flow in this greenhouse tunnel is controlled by three reversible flow fans at one end of this tunnel, and other fans control the air flow rate and air residence time to maintain the air temperature gradients from one end to the other of the greenhouse tunnel. During the night the directions of fans are reversed, and heat is added to the system with a burner to maintain a similar gradient of air temperature with respect to ambient air temperature. This system is used to track ambient air temperature conditions, with the warmer end of the

D.C. Uprety, V.R. Reddy, *Crop Responses to Global Warming*,
DOI 10.1007/978-981-10-2004-9_3

tunnel maintained at a constantly higher temperature than the cooler end of the tunnel. Consequently, plants can be exposed to a range of increased temperatures as a result of the gradient of increasing temperature that is established in the chamber—hence the name temperature gradient chamber (TGC). Overhead fans are used to prevent the development of vertical temperature gradients. Air temperature is sensed throughout the tunnel.

Advantages:

1. Air flow for maintaining the temperature gradients is sufficient to prevent any measurable variation between modules.
2. This facility is computer controlled.

Disadvantages:

1. The small size and low-light condition limit the usefulness of this facility for meaningful studies on crop responses to rising temperature.
2. The controlled air flow rate and air residence time experienced by the crops are highly unnatural.
3. It is a complicated system which needs continuous technical attention.
4. It has unnatural solar and long-wave radiation environment and air flow above the canopy, which is not closer to the field conditions (Fig. 3.1).

Schematic drawing of the temperature gradient chamber (TGC) for CO_2 enrichment (bottom) and ambient CO_2 (top) developed at Kyoto, Japan. (adapted from Horie *et al.*, 1995) 1,oil heater; 2,stovepipe of oil heater; 3,air exhaust window; 4,variable speed exhaust fan; 5,reversible exhaust fan; 6,oscillating fan; 7, CO_2 controller; 8, CO_2 injection pipe; 9,liquid CO_2 tanks.

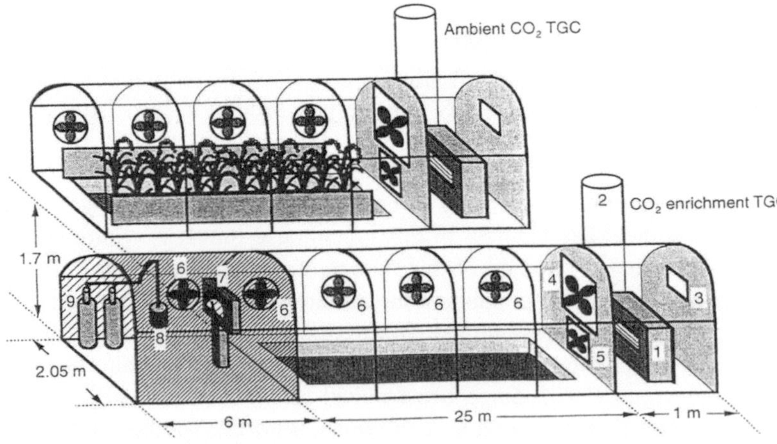

Fig. 3.1 Temperature gradient chamber (Horie et al. 1995)

Rawson (1995) designed 2.45 m long, 1.25 m wide, and 1.25 m high-temperature gradient chamber to grow wheat plants for studying the response of different cultivars to variable temperature and CO_2 concentration. These are covered with 6 mm thick UV stable polythene sheet. This film is elastic and reduces the damage of water and hail. Its outlet plenum is a box 0.65 m in length, 1.25 m width, and 1.25 m height covered with PVC sheet. The fan is mounted pointing toward the roof. The temperature difference between inlet and outlet does not exceed more than 5 °C. Temperature measured by differential thermocouples is used to set the required fan speed electronically. Air flow needed to maintain temperature gradient is sufficient to prevent any measurable difference. A temperature gradient tunnel was developed as an indoor plant growth cabinet facility (Grime 1989). This system can quickly provide sets of temperature responses for morphologically smaller plants. The facility is computer controlled and can create a temperature gradient over the range of 10–35 °C and also inject steam for humidification. However, the levels of photosynthetically active radiation were only about 125 umol (photons) m^2s^1. The small size and low-light conditions limit the usefulness of such a facility for meaningful temperature effect research (Fig. 3.2).

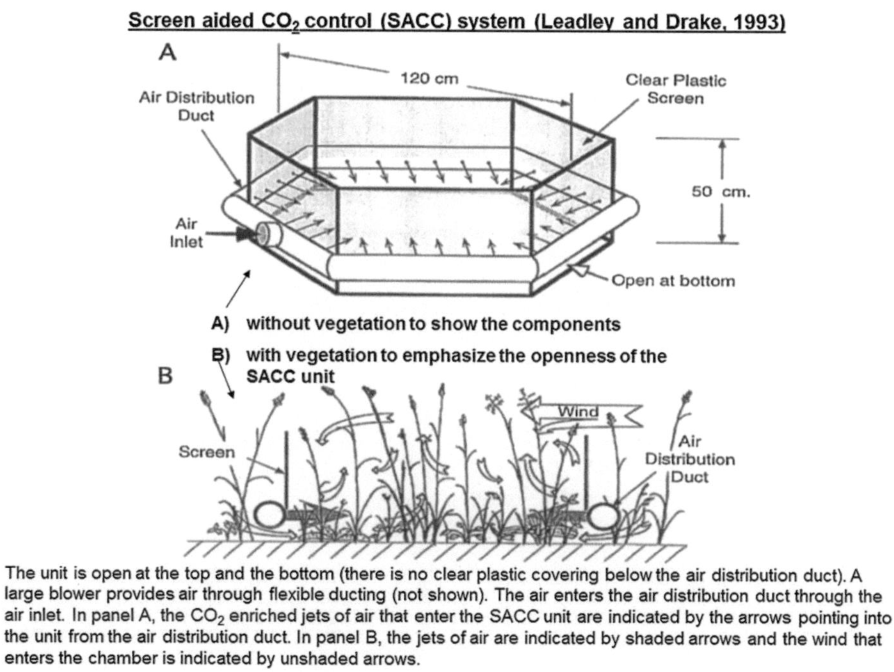

Fig. 3.2 Screen-aided CO_2 control (SACC) system (Leadley and Drake 1993)

3.2 Temperature Gradient Greenhouses (TGG)

Each temperature gradient greenhouse is built on a semicircular galvanized steel framework and covered with six light polythene terephthalate films, which transmit about 90 % of the incoming solar photosynthetically active radiations. Each TGG is 27.4 m long, 4.5 m wide, and 2.2 m high at the apex. Each greenhouse is maintained by regulating the rate of unidirectional ventilation relative to the energy input of heated air from natural gas heaters located outside the TGG and incident solar radiations (during the daytime). At night and during low solar irradiance periods of the day, heated air is introduced via ducts continuously to a point with the overhead paddle from near the top of TGG. Air temperature is measured in the center of each section with shielded and aspirated copper constantan thermocouples positioned 0.6 m above the soil surface. CO_2 concentration is maintained at ambient level throughout the TGG. PAR is measured with a calibrated quantum sensor. Data from temperature and radiation sensors are recorded every minute. Moving averages at hourly intervals over 24-h period are being taken for calculation purpose using mean daytime and mean nighttime averages (Fritschi et al. 1999) (Fig. 3.3).

3.3 SPAR System (Soil Plant Atmosphere Research System)

Soil Plant Atmosphere Research facility has been modified to study the response of crop plants to increased temperature. It is generally located outdoors on a 20 × 30 m concrete pad. Each SPAR unit has the capability of controlling air temperatures and atmospheric composition at predetermined set points. Controlled factors in

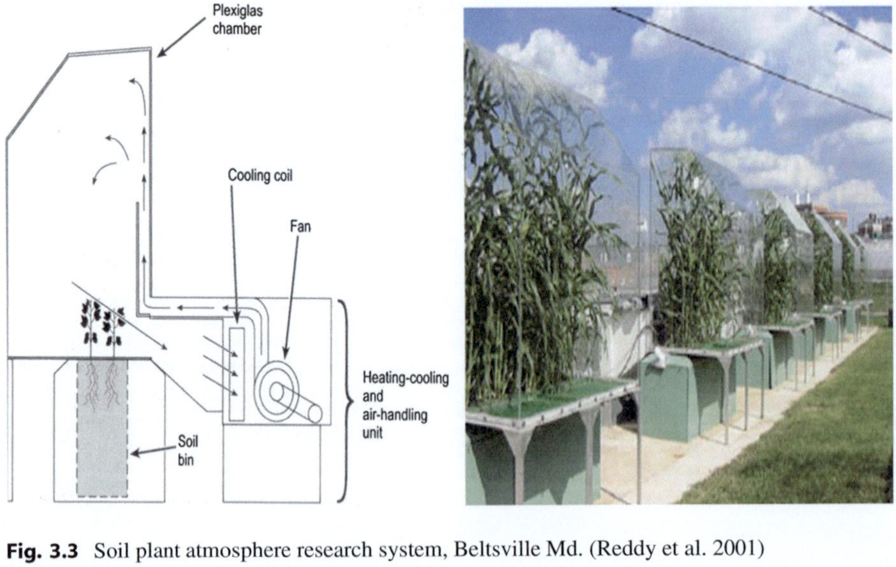

Fig. 3.3 Soil plant atmosphere research system, Beltsville Md. (Reddy et al. 2001)

each chamber include atmospheric CO_2 concentration, dry bulb air temperature, and dew-point temperature. Temperature is monitored and adjusted automatically, every 10 s throughout the day and night. Control of the dry bulb air temperature is maintained using a computer that opens and closes a set of solenoid valves to a chilled water radiator and switches a heating system on and off. Heat is provided by two 11 kW heating elements mounted on either side of the air circular unit. Air temperature is monitored using thermocouple maintained within +/− 5 °C of the treatment set points over a daytime range of 18–40 °C and a nighttime range of 12–32 °C. Canopy temperature is monitored after every 10 s using an infrared thermometer (Reddy et al. 2001).

3.4 Infrared (IR) Warming Technology

Kimball (2005) designed and developed T-FACE facility to warm 3.0 m diameter surface area with IR warming system. Target set point of 1.5 °C during daytime and 3.0 °C during nighttime more than ambient temperature was used to study the temperature-induced changes in crop plants. Infrared heaters were deployed in and mounted in reflected housing in hexagonal arrays. Heaters were suspended from steel cables in a triangular tripod configuration at a height of 1.2 m above the wheat canopy and were deployed at an angle of 45° from horizontal and 30° with respect to steel triangular suspension tripod configuration. Heaters were raised weekly as the crop grows. Canopy temperature was measured using IR thermometers. Before deployment in the field, the IR thermometer was calibrated far and wide range of temperature (−5 to +70 °C). Infrared thermometer was kept 0.30 m above crop canopy and pointed them with an oblique angle of 30° below horizontal. The IR warming with a T-FACE apparatus was an effective way to investigate the impact of global warming on an agroecosystem such as field-grown wheat. Uprety and Reddy (2008) developed such temperature enrichment facility to study the interactive effect of CO_2 and temperature in crop plants under South Asian conditions. The analysis of days to anthesis and duration of grain filling indicated that T-FACE is an effective technique for providing warming treatments that mimic natural temperature variations. The infrared warming and automatic control of temperature rise have shown potential as a procedure for warming field plots above ambient temperatures avoiding the effects of altered solar radiations, precipitation, and wind. According to White et al. (2011), T-FACE induced no artifacts, notwithstanding the feasibility of the IR warming system with PID control of constant temperature rise used in conjunction with supplemental irrigation. This successful deployment of the system in field condition was first described by Kimball et al. (2008) (Fig. 3.4).

3.5 Free Air Temperature Enrichment Technology (FATE)

This technology is developed to simulate the anthropogenic-induced climate warming. The components associated with FATE technology include two 1500 W infrared lamps positioned at about 1.2 m above the ground, which can uniformly irradiate

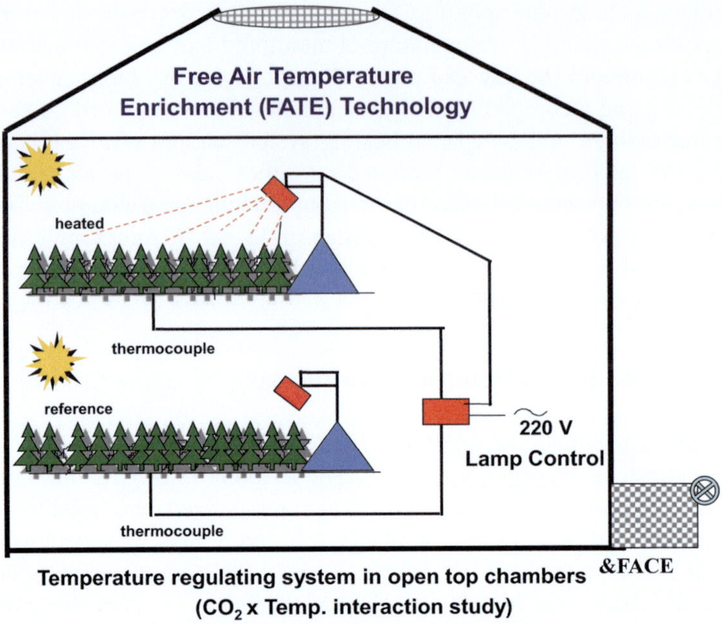

Fig. 3.4 Free Air Temperature Enrichment Technology (FATE) proposed to be added with OTC and FACE facilities operating under South Asian Conditions (Uprety and Reddy 2008)

40–50 cm path of vegetation at an angle of 40° aboveground. These lamps are regulated by proportional action controller, which modulates the IR flux density at a frequency of 10 Hz to obtain a preset target differential (T) between heated and a corresponding unheated plot. The infrared lamps used in simulating the temperature are made of tungsten filament and can irradiate at a temperature of approximately 2000 °C. A filter is used to selectively cut off visible light so that light can be continued during night. It also removes the wavelength to which photochromes are sensitive so that photomorphogenesis can be avoided. The FATE radiations are uniform in space and time. NijsI et al. (1997) modified this FATE facility to irradiate larger area with an array of six lamps, each with their individual waterproof housing, positioned about 1.5 m above the ground. Each lamp house has fans to remove extra heat, rain shield, and two radiation shields to obstruct any infrared light reflection from rain shield and filter to cut off visible light (Fig. 3.5).

Kimball (2005) designed and developed a proportional integrative derivative control system for controlling the rise in temperature. The degree of warming was maintained using PID control system with Apogee infrared radiometers to sense the canopy temperatures of the heated plots as well as the controlled one. Arrays of infrared heaters can be used to warm open-field plots and thereby enable the interactions of increasing CO_2 and higher temperature to be studied under free-air conditions. This experimental approach has been called T-FACE (temperature free-air controlled enhancement). T-FACE consists of an array of six infrared heaters

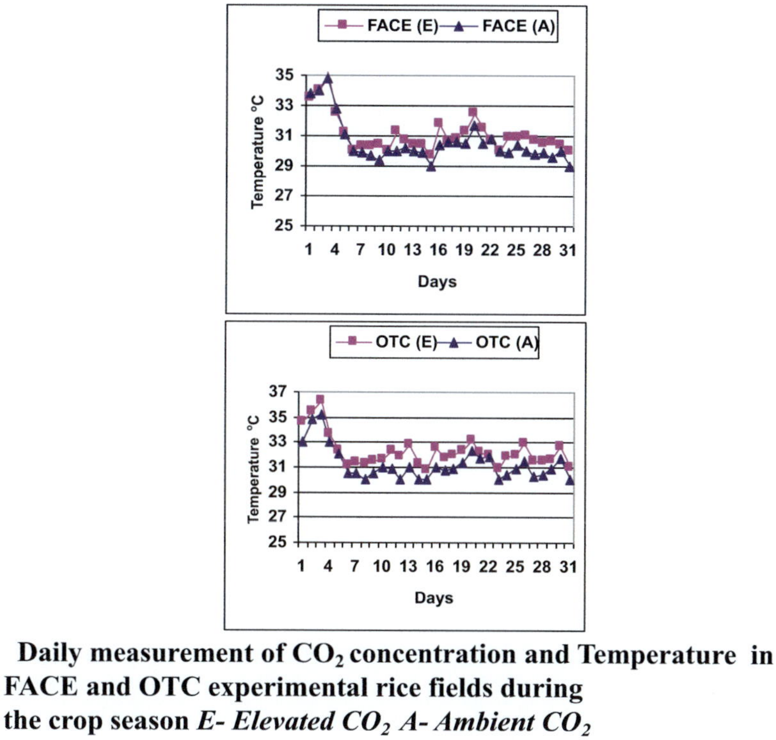

Fig. 3.5 Comparison of temperature and CO_2 concentration in FACE vs. open to chambers (Uprety et al. 2006)

placed above the crops in a hexagonal pattern for uniformly increasing the temperature of a 3 m diameter circular field area. The shading effect of this instrument on field crop is less than 1 %. The potential effect of any other environmental factor, if any, could be minimized by comparing high-temperature plots with controlled plots having dummy nonoperative heaters. According to Ottman et al. (2012), this technique can increase the temperature of the experimental open-field plots without the compounding effects of other environmental factors. Dejonge et al. (2015) suggested infrared thermal radiometers (IRT) as an affordable tool to monitor canopy temperature. They have identified temperature measurements on degrees above canopy threshold (DACT) as the most convenient index to understand the response of maize plants to the higher temperature. It requires a single canopy temperature measurement, which has a strong relationship with the parameters associated with the response of the crops like *Zea mays* to high temperature.

It is important that these technologies demonstrate convincingly the extent of inherent variability in their ability to control and regulate temperature and to ensure a high level of operational integrity. With further engineering development and testing, more realistic and cost-effective methods for rising atmospheric temperature experiments in open-air setting will be designed. These technologies present a window into likely future of ecosystem function under warming conditions.

The focus of the individual experiments tended to be directed either at soil processes or plant processes. Thus more comprehensive, whole ecosystem studies are needed in order to test the validity of both conceptual and quantitative models of an ecosystem to warming. Following suggestions will be useful for planning the future research on ecosystem response to global warming:

1. Focus on the whole ecosystem warming experiments.
2. Support long-term experiments.
3. Add experiments in underrepresented biomes.
4. Encourage common protocols.
5. Combine warming studies with temperature gradients.
6. Investigate interactions with other drivers of environmental changes.

3.6 Soil Warming System

Siebold and Von Tiedemann (2012) developed a robust soil warming facility in Germany to study the impact of higher soil temperature on crops. It is comprised of heating cable engraved in the field with a computerized temperature control and data logging system. Heating was regulated for medium- and long-term climate warming scenarios reflecting the duration of the period up to 2050 and 2100, respectively, in comparison to the ambient temperature control. They have programmed temperature stable at two levels, i.e., +1.6 °C (±0.17 °C) or +3.2 °C (±0.27°C) above ambient temperature throughout the crop duration. They have studied the effect of soil warming on the field microclimate as well as on biological characters in winter oilseed rape (*Brassica napus*). This facility was developed at Heppenheim, Germany. The facility includes plastic containers (116 cm × 80 cm × 60 cm) filled with compost soil. Warming of the soil was done using a heating cable with 30 W m^{-1} reaching 300 W m^{-2} at 5 cm spacing. The temperature control using universal thermostat along with a sensor set with the target temperature of 11 °C was placed in the middle of the experimental area. The soil temperature was monitored at different places in the container using temperature probes and data logger. Soil water content was measured and kept at –100 to –150 Kpa using a water mark sensor connected to data reader. The variation of setting included cable spacing (5 cm and 10 cm), cable depth (5,10,15, and 20 cm), and placement of the temperature control thermistor (adjacent to cable, between the cables, 5 cm above cables, and 5 cm below cables). The setup was tested for 24–36 h and calibrated properly before it was used for experimental purposes.

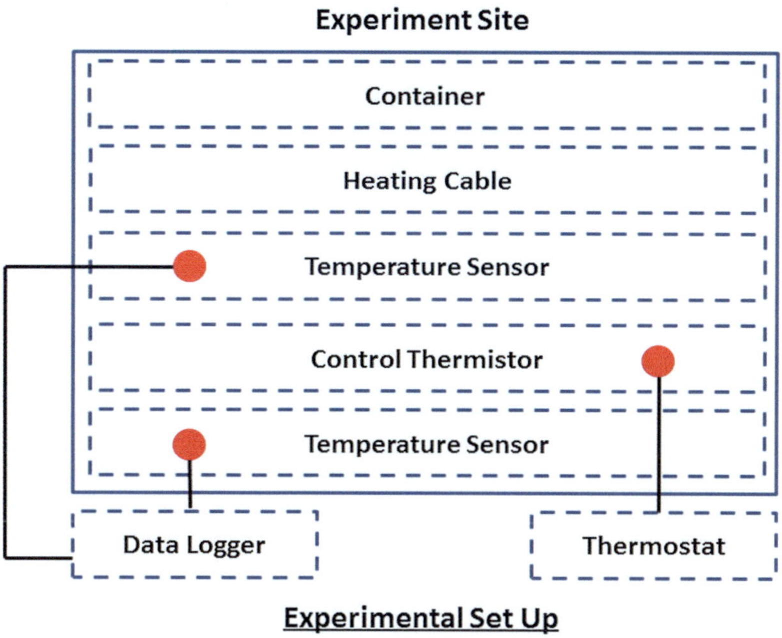

Siebold and Von Tiedemann (2012)

References

Dejonge KC, Taghvaeian S, Trout TJ et al (2015) Comparison of canopy temperature-based water stress indices for maize. Agric Water Manag 156:51–62

Fritschi FB, Boote KJ, Sollenberger L (1999) Carbon dioxide and temperature effects on forage establishment: photosynthesis and biomass production. Glob Change Biol 5:441–453

Grime JP (1989) Ecological effects of climate change on plant populations and vegetation composition with particular reference to British flora. In: Jackson M et al (eds) Climate change and plant genetic resources. Belhaven press, London, pp 40–60

Horie T, Nakano J, Nakagawa H et al (1991) Effect of elevated CO_2 and high temperature on growth and yield of rice. 1. Development of temperature gradient tunnels. Jpn J Crop Sci 60:127–128

Horie T, Nakagawa H, Nakano J et al (1995) Temperature gradient chambers for research on global environmental change. III.A system designed for rice in Kyoto, Japan. Plant Cell Environ 18:1064–1069

Kimball BA (2005) Theory and performance of an infrared heater for ecosystem warming. Glob Change Biol 11:2041–2056

Kimball BA, Conley MM, Wang S et al (2008) Infrared heater arrays for warming ecosystem field plots. Glob Change Biol 14:309–320

Leadley PW, Drake BG (1993) Open top chambers for exposing plant canopies to elevated CO2 concentration and for measuring net gas exchange. Vegetation 104(105):3–15

NijsI FR, Blume H et al (1997) Stomatal regulation in a changing climate: afield study using free air temperature increase(FATI) and free air CO_2 enrichment (FACE). Plant Cell Environ 20:1041–1050

Ottman MJ, Kimball BA, White JW et al (2012) Wheat growth response to increased temperature from varied planting dates and supplemental infra red heating. Agron J 104:7–16
Rawson HM (1995) Response of two wheat genotypes to carbon dioxide and temperature in field studies using temperature gradient tunnels. Aust J Plant Physiol 22:23–32
Reddy KR, Read JJ, Baker JT et al (2001) Soil-Plant Atmosphere-Research (SPAR) facility – a tool for plant research and modeling. Biotronics 30:27–50
Siebold M, Von Tiedemann A (2012) Potential effects of global warming on oil seed rape pathogens in northern Germany. Fungal Ecol 5:62–72
Uprety DC, Garg SC, Bisht BS et al (2006) Carbon dioxide enrichment technologies for crop response studies. J Sci Ind Res 65(11):859–866
Uprety DC, Reddy VR (2008) Rising atmospheric CO_2 and crops. ICAR Publications, New Delhi, pp 1–144
White JW, Kimball BA, Wall GW et al (2011) Response of time of anthesis and maturity to sowing dates and infrared warming in spring wheat. Field Crop Res 124:213–222

Case Histories: Crops

4

4.1 Wheat

Wheat is the most important crop for the human and animal nutrition. It is highly vulnerable to the global rise in temperature adversely affecting its production and consequently the human food supplies. Wheat production affects the health, nutrition, and livelihood of the world's large population since its area of production is distributed in the larger part of the world (FAO 2010). IPCC prediction indicates that wheat production will decrease substantially. The modeling studies predict that wheat will be grown at higher latitudes and the production in regions closer to the equator will decrease (Leemans and Solomon 1993). The grain yield in wheat at 2 °C warming might improve by 12 % due to longer crop duration, but a 4 °C warming would decrease global production. Narayanan et al. (2015) considered a high temperature as the major environmental factor that limits the wheat productivity.

Wheat crop in India is grown in varied agroclimatic conditions from 15° North to 32° North and from 72° East to 92° East. It experiences different temperatures during the crop season depending upon the growing place. The mean temperature during the crop season is 16 °C in the northern region, 20 °C in eastern region, and 23 °C in central and peninsular India. The wheat growing period in India, from sowing to harvesting, is between October–November and March–April. The sowing of wheat is regulated by the gradual cooling of the seed bed. The ripening and harvesting are controlled by the subsequent rise in temperature during the month of March–April. Thus, from sowing to harvesting, the wheat growing period is temperature limited, and it restricts the choice of early maturing varieties and their geographical distribution. The wheat growing period becomes shorter from the north to south as the duration of winter decreases. The grain yield of wheat in the southern and eastern region of India is comparatively lower. Howard (1924) pointed out that "wheat growing in India is a gamble in temperature." Wheat in India is invariably exposed to very high temperatures during some stages of development.

D.C. Uprety, V.R. Reddy, *Crop Responses to Global Warming*,
DOI 10.1007/978-981-10-2004-9_4

Table 4.1 Growing season temperature, crop duration, and yield in wheat at two locations in India (1984–1985) (Data based on 16 varieties)

Duration and severity of winter decreases from north to south; correspondingly growing season becomes shorter and productivity declines			
Crop duration (Days)	Yield (Q.ha^{-1})	Mean temperature (°C)	Places
127	45.2	16.9	Delhi
108	32.8	20.0	Powerkheda

Growing period extends as much as 7–8 months in northern hills where temperature is low
Source: AICWIP Annual report

All stages of development are sensitive to temperature. According to Long and Ort (2010), the rising temperature may make the Indo-Gangetic plains of India from a high potential wheat production area to a heat-stressed, short-season wheat-producing area. Indo-Gangetic plains produce approximately 15 % of the world's wheat crop. The identification and development of high-temperature stress-tolerant wheat varieties will be critical for the food security of this South Asian region and Indian subcontinent (Table 4.1).

Wheat is grown in India under a wide range of soil and climatic conditions as well as in the hills up to the altitude of 10,000 feet above sea level. The sowing is determined by gradual cooling of the seed bed after slowly receding warmer period for an initial seedling establishment. Maturation has to be completed before rapidly ascending temperatures combined with hot and dry winds, which starts in summer. The planting of the crop around mid-November is the established practice under irrigated conditions. However, seeding is advanced by about 4 weeks in unirrigated areas so that the crop is able to utilize the moisture stored during the preceding monsoon. The ripening of the crop terminates during March–April in Punjab, Haryana, Uttar Pradesh, Bihar, Madhya Pradesh, and Rajasthan. Duration and severity of winter decrease from north to south. The growing season becomes shorter and productivity declines, whereas the winter is warmer and the yield is low in the eastern region (Odisha, Assam, and West Bengal). The All India Coordinated Wheat Improvement Project data analysis indicates that the reduction in grain yield corresponds with the increase in temperature. Crop duration and yields were reduced from 127 days and 45.2 q h^{-1} (mean temperature 16.9 °C at Delhi) to 108 days and 32.8 q h^{-1} (mean temperature 20 °C at Powerkheda). The growing period increases as high as 7–8 months in northern hills where the temperatures are low. The farmers of the northern region of high elevation prefer to grow a summer crop in May and June and harvest in September/October. Under such situations, sowing of wheat is delayed toward the last week of November to the end of December or even the beginning of January, and consequently later stages of the crop are pushed into rising temperatures in ensuing months. This causes a reduction of 4 months in the growth period. However, wheat growing area will likely shift northward as a result of global warming. Plant breeding and biotechnological methods may help in developing wheat varieties tolerant to high temperatures of central and southern part of India.

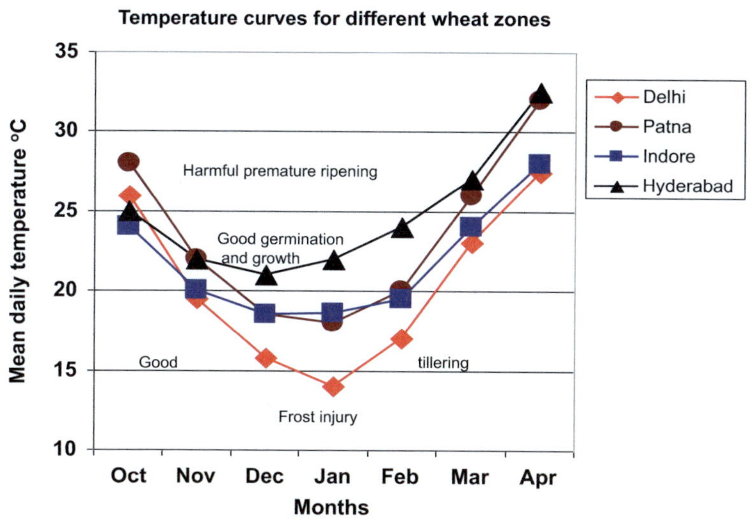

Fig. 4.1 Change in temperature during wheat growing season in different wheat zones (Adapted from Bhardwaj 1978)

Ottman et al. (2012) in an exclusive experiment using T-FACE technology in the USA demonstrated that supplemental heating decreased grain yield of wheat by 6.9 % per 1 °C increase above optimum temperature. The planting of wheat in winter had no effect of supplemental heating on grain yield. However, supplemental heating in spring planting decreased the grain yield from 510 to 368 g m^{-2}. This reduction was due to reduced green leaf area, reduction in the number of productive tillers per plant, reduced grain weight, fewer grain numbers per spike, and decreased harvest index. They concluded that the future expected global warming may reduce grain yields of late planting wheat and may require a shift for early planting in arid regions of the Southwest USA (Fig. 4.1).

4.1.1 Growth and Development

The rates of development of wheat are sigmoid rather than linear functions of temperature. Development starts above a base temperature between 1 and 5 °C depending on the type of cultivar. The rate of development increases slowly as temperature rises followed by a rapid development with maximum intensity at 30 °C. These studies indicate that a 1 °C warming would reduce crop duration by about 21 days (8 %) and the reproductive period by about 8 days (6 %). The duration from anthesis to maturity is shortened by about 3 days and the grain-filling period by about 2 days (5 %) per 1 °C increase in temperature (Shaykewich 1995; Mitchell et al. 1995). The biomass production in general decreases with increase in temperature. However, the adverse effect of the rise in temperature was greater on grain yield than that of biomass. The shortening of the maximum growth rate period will have more effect on final biomass than the shortening of early growth stages by proportionately reducing the harvest index and grain yield. Mitchell et al. (1995) in this study also demonstrated that an increase of 3.5 °C temperature throughout the growing season reduced the production of biomass by 16 % and grain yield by 35 %.

Wheat experiences heat stress of varying degrees at different phenological stages; however, the heat stress during the reproductive phase is more harmful than a vegetative stage, due to its direct effect on grain number and dry weight (Wolenweber et al. 2003). Heat stress during reproductive phase can cause pollen sterility, tissue dehydration, reduction in photosynthesis, and increased photorespiration. When the temperature is increased between anthesis and grain maturity, grain yield is reduced because of short active duration for grain formation and grain filling. Lobell et al. (2005) in Mexico demonstrated a reduction of 10 % of wheat grain yield for every 1 °C increase in night temperature, but the same increase in temperature during daytime had no significant effect. Increased night temperature linearly decrease the duration of grain filling. Prasad et al. (2008) found that night temperatures of 20 and 23 °C reduce grain-filling period by 3 and 7 days, respectively. Narayanan et al. (2015) showed that high night and day temperature increased the concentration of reactive oxygen species (ROS) and membrane damage with reduction in the antioxidant capacity and photochemical efficiency. They have observed that the combination of day and night high temperature was more deleterious than their independent effects.

The developmental stages of the crop exposed to increased temperature have a variable effect on the damage experienced by the plants (Stafer and Rawson 1995). The high temperature at double ridge stage causes less damage to the productivity of wheat, whereas high temperature at anthesis and subsequent grain development stages cause severe damage in the productivity of wheat. The major yield components reduced by the increase in temperature are the harvest index. The harvest index was lower in plants experiencing higher temperature at the reproductive stage because their grain number per plant was reduced drastically. Development generally accelerates linearly as the daily average temperature increases. Growing degree days showed the similar trend to a linear model of developmental response to temperature. Such linear model describes wheat development as long as temperature remains within 10–30 °C. However, a nonlinear model is required to explain development under extreme temperature stress. Phenological stages differ in sensitivity to temperature. The effect of temperature on the duration of these phenological stages has been studied on the basis of the accumulation of heat units referred as growing degree days (GDD). This is estimated as a cumulative mean diurnal temperature above a base value of 4.5–5.0 °C required to complete a specific phenological event. It is well documented that a given variety has a definite temperature (GDD) and photothermal requirement before it attains a certain phenological stage in any environment. Thus, an increase in temperature will not disturb the plant GDD but cause a reduction in the duration of the phase (calendar time) (Rawson 1988). Chakravarty and Sastri (1983) observed almost similar GDD from sowing to anthesis for a wheat cultivar Kalyansona, i.e., 1086 and 1069 to November sowing in warmer and cooler seasons, respectively, but the time of anthesis was reduced by 9 days in warmer conditions. Delayed sowing, however, reduced the time of maturation of the crop. The Kalyansona variety of wheat accumulates more or less similar units of GDD (almost 950) in three temperature regimes (15/10, 21/16, 27/22 maximum/minimum) under controlled condition (Rawson and Bagga 1979). The duration to anthesis was reduced by 25 and 35 days, respectively, in later two treatments (Fig. 4.2).

4.1.2 Physiological Responses

Physiological responses to temperature changes in plants may occur at short and long time scales (Wollenweber et al. 2003). Short-term effects involve gene expression, such as heat shock protein synthesis which is likely to occur. Long-term responses include alterations in the rate of carbon assimilation, electron transport per unit leaf area and impaired cell anaplerotic carbon metabolism, sucrose synthesis, and C-N ratio between the organs resulting in to the changes in growth rate and grain yield. Lawlor and Mitchell (2000) showed that net photosynthesis is reduced (25 % of maximum) at 5 °C and increases as temperature rises more than 25 °C and declines at about 40 °C. This is due to kinetic parameters of Rubisco and relative solubility of CO_2 and O_2, which mean that photorespiration increases with temperature.

The short-duration temperature changes markedly affect the respiration in wheat crop, whereas the sensitivity of wheat respiration is much less in long-term temperature changes. It was contended that among growth and maintenance respirations, the latter being highly temperature dependent (Gifford 1995). Heat stress reduces the net leaf photosynthesis through disruption of structure and function of chlorophyll content (Xu et al. 1995). Photosystem II (PS II) is often considered to be highly vulnerable to heat stress (Enami et al. 1994); however, Yamane et al. (1998) demonstrated that damage to PS II only occurs at the temperature above 45 °C. The decrease in photosynthesis under moderate heat stress occurs through a decrease in CO_2 assimilation (Long et al. 2004). The loss due to the rise in photorespiration with an increase in temperature makes CO_2 unavailable to Rubisco for carboxylation. Salvucci and Crafts-Brandner (2004) showed that Rubisco activase is significantly vulnerable to high temperature and it is considered the major limitation to the carbon fixation under moderately high temperature. High-temperature stress denatures Rubisco activase and this inactive Rubisco is not converted to active form. Salvucci et al. (2001) reported denaturation of Rubisco activase at temperature as low as 32 °C. The presence and the incorporation of thermostable Rubisco activase into crop plants will help in maintaining Rubisco in active state even under high temperature. Kurek et al. (2007) using gene shuffling technology provided clear evidence that Rubisco activase is a major limiting factor in plant photosynthesis under moderately higher temperature and a potential target for genetic manipulation to improve the productivity of wheat under higher-temperature stress.

Phenological events of wheat plants

1. Germination
2. Double ridge or first spikelet initiation
3. Terminal spikelet initiation
4. Ear emergence
5. Anthesis and
6. Grain development

1 wheat-seedling

2 wheat-Tillering

Fig. 4.2 Photographs of wheat plants at different phonological stages

3 wheat-flowering

4 wheat-Panicle initiation

Fig. 4.2 (continued)

5 wheat-Grain filling

6 wheat-Maturation

Fig. 4.2 (continued)

Cure and Acock (1986) suggested that wheat yield may increase by 35–40 % by doubling CO_2. It is likely that temperature may rise to as much as 4 °C at this level of CO_2. The rise in atmospheric CO_2 is expected to reduce the negative effect of temperature on yield by enhancing the rate of photosynthesis and the efficient translocation of this additional photosynthate to developing grains (Wheeler et al. 1996a, b). An increase of 17 % in wheat grain yield per 100 umol mol^{-1} of CO_2 enrichment has been reported by Batts et al. (1997), but wheat grown at warmer temperature (inside poly tunnel) showed a reduced harvest index (Wheeler et al. 1996a, b). These reports suggest that grain yield is more sensitive to temperature than biomass. However, the elevated CO_2 may mitigate the adverse effect of high temperature on wheat grain yield.

The increase in the concentration of CO_2 at the sight of Rubisco is another mitigation technique to counter the adverse effect of high-temperature stress as adopted by C4 plants. This reduces the photorespiration and decreases the carbon loss by reducing the oxygenation reaction. This strategy of C4 plants makes them adapted to high-temperature stress and dry regions. Several efforts are being made to incorporate C4 type of photosynthesis into C3 plants (Leegood 2002). However, such an effort requires not only C4 photosynthesis cycle but also the Kranz leaf anatomy and expression of different photosynthetic proteins (Zhu et al. 2010). Inactivation of chloroplast enzymes due to higher temperature induces oxidative stress and reduces photosynthesis. Sairam et al. (2000) observed that this oxidative stress induces lipid peroxidation leading to protein degradation, membrane rupture, and enzyme inactivation. Salviecci and Crafts-Brandner (2004) reported that heat stress for 24 h in dark irreversibly decreased the Rubisco subunits and Rubisco activase. The Rubisco is more sensitive to increased temperature compared to other enzymes of carboxylation process. Although Rubisco catalytic activity increases with temperature, its low affinity for CO_2 limits the chance of increasing net photosynthesis with temperature. The solubility of O_2 decreases to a lesser extent than CO_2 resulting in increased photorespiration and lowers photosynthesis (Lea and Leegood 1999). Heat stress and excessive light may damage PS ll active sites due to the fluidity of thylakoid membrane and electron transport-dependent integrity of PS ll and cessation of photophosphorylation (Prasad et al. 2008). Heat stress-induced breakdown of thylakoid membrane may also result into leaf senescence.

The phenological stage from sowing to 1st spikelet initiation is less sensitive to temperature. The greatest effect of the rise in temperature on yield is from spikelet initiation to terminal spikelet formation until anthesis. Spikelet number and floret numbers, both the major parameters of yield, establishes during this phase. Grain weight appears to be much less sensitive to heat stress than is grain number. Asana and Williams (1965) reported 16 % reduction in grain weight for each 5 °C rise in temperature. Every 1 °C rise in temperature, there is a depression of 8–10 % in grain yield mediated by 5–6 % fewer grains and 3–4 % reduction in grain weight. According to Wardlaw (1974), this reduced grain filling was attributed to:

1. Source—flag leaf blade
2. Sink—ear
3. Transport pathway—peduncle

Thus, high-temperature-induced reduction in grain yield was attributed to:

1. Reduced number of grains formed
2. Shorter grain growth duration
3. Inhibition of sucrose assimilation in grains

He observed that photosynthesis had a broad temperature optimum from 20 to 30 °C. Accordingly the photosynthesis of flag leaf declines rapidly above 30 °C. The rate of ^{14}C assimilate movement out of flag leaf to phloem loading was optimum around 30 °C, whereas the rate of ^{14}C assimilate movement through the stem was independent of temperature up to 50 °C. This demonstrates the effect of temperature stress on both source and sink activity in wheat plants. Asana and Saini (1962), Asana and Joseph (1964), and Asana and Williams(1965) reported 13.9 % and 17.8 % reduction in the grain weight of wheat varieties PbC281 and NP720 for an increase of 5 °C in the average temperature. There was not much variation among the cultivars for this character indicating that the grain size is a stable character. However, senescence is generally hastened slightly by high temperature. Thus, early varieties can escape the deleterious effect of high temperature during grain filling.

Wheat growing in India experiences both temperature stress and moisture stress simultaneously, which adversely affects the wheat productivity. Bharadwaj (1978) in a study on temperature calendar for wheat growing suggested a range of 20–25 °C temperature favorable for germination and seedling emergence, whereas 16–20 °C good for tillering. Thus, mid-November temperature is conducive for desirable tillering in irrigated region. In North India, early medium varieties (90–100 days flowering) show tiller number at its peak 50–55 days after sowing and subsequently decline to a constant number (Nigam 1977). Low temperature is beneficial for wheat crop results in better tillering. The rise in temperature can change this time sequence and tillering behavior. Both mid-October and late December–January planting will suffer more for the need of adequate tillering because the increase in temperature during warmer months of November and rising hotter conditions from late January onward coincides, respectively, with their tillering phase reducing its duration and number. This will reduce the growth rate, accelerate heading, and depress the plant growth. The rise in temperature in these phenological stages will prove drastic for the germination and tillering. The terminal stress badly affects the wheat productivity especially in late sown wheat, which is a common practice in a cotton-wheat cropping system. Changes in daily maximum temperature during flowering and grain filling adversely affect the grain yield. During grain filling a significant negative impact of heat stress was reported in South Australian wheat through the greatest decline in fertility. However, there was less adverse effect of heat stress on leaf characters. But leaf senescence was increased if heat stress occurs at post-flowering and early grain-filling stages (Telfer et al. 2013).

Heat stress in the rhizosphere occurs when air temperature exceeds more than 10–15 °C in barren and dry soil and the maximum temperature of soil reaches from 40 to 45 °C with serious effects on seedling emergence. Patil et al. (2010)

demonstrated that soil warming hastened crop development at early stages till stem elongation. Soil warming also increased green leaf area (GLA) without affecting the production of tillers. However, it reduced the number of ears and increased the grain weight. Siddique et al. (2000) reported a mean reduction of 4 % in grain weight per degree increase in mean temperature during grain filling. Crop growth mean temperature increase from 12.2 to 27.5 °C caused 34.5 % reduction in grain filling per day. The high temperature is most detrimental for wheat crop during grain developmental phase in the month of March. Temperature out of optimum cardinal temperature, i.e., both upper and lower limit, is detrimental depending upon the stage and type of the crop. Prasad and Djanaguiraman (2014) demonstrated that the rise in temperature before anthesis and at reproductive developmental stages were highly detrimental for the wheat crop. They also observed a linear reduction in floret fertility and grain mass with increase in the period of high-temperature stress. The high-temperature-induced reduction in floret fertility was attributed to abnormalities in pollen, stigma, and style. The reduction in floret fertility and individual grain weight leads to a greater reduction in wheat grain yield. Heat stress during anthesis also increases floret abortion. Wheeler et al. (1996a, b) reported that higher-temperature stress between flowering and grain development caused a significant reduction in grain yield and attributed it to the shorter duration for the transfer of photosynthates to grain development. Grain number and grain size can be substantially reduced if susceptible cultivars are exposed to high-temperature stress at flowering. Temperature above 31 °C after anthesis and grain filling reduces the net assimilation causing the production of small grains and increasing leaf senescence. Savin et al. (1999) reported that the high-temperature stress at pre- as well as post-anthesis stages significantly diminishes the grain weight and their effect was similar with the exception of heavier grains which remains unresponsive to the high-temperature stress at pre-anthesis stage. Although the rate of grain growth increases with the increase in temperature, this increases the grain-filling rate compensating for the shorter grain-filling period. However, this compensation does not occur at temperatures above 30 °C due to a reduction in the translocation of assimilates from flag leaf to grain. The threshold value of temperature for heat stress effect was 31 °C for flowering and 35 °C for grain filling. Yield reduction was associated with temperature higher than 32 °C during mid and late reproductive stages including grain filling (Fig. 4.3).

Higher temperature reduces the conversion of sucrose to starch in developing wheat endosperm (Bhullar and Jenner 1986). The accumulation of starch was more vulnerable than an accumulation of protein (Bhullar and Jenner 1985). The control of starch synthesis in wheat endosperm lies with the activity of soluble starch synthase, which is highly sensitive to high-temperature stress (Jenner et al. 1995). Zahedi et al. (2003) demonstrated that the tolerance for higher temperature during grain development in wheat cultivars depends on their efficiency of soluble starch synthase activity. The role of low molecular weight heat shock proteins located in plastids of heat-tolerant wheat cultivars attributed for providing thermotolerance to grain growth and protection to soluble starch synthase enzyme, which is sensitive

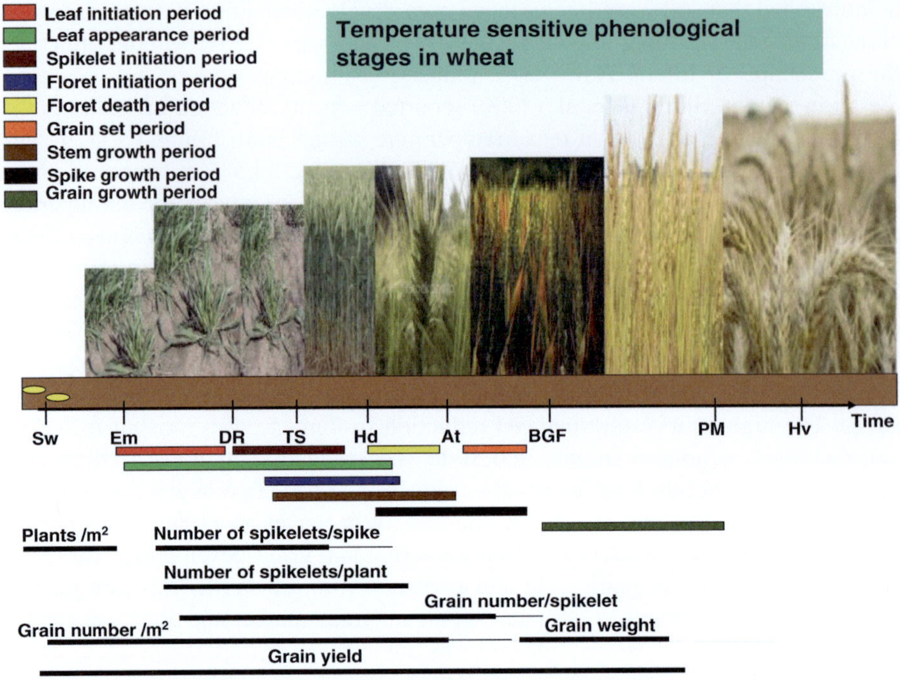

Fig. 4.3 Wheat phenophases (Modified from Slafer and Rawson 1994)

to high-temperature stress. Thus, thermotolerance for grain growth in wheat can be improved by the incorporation of the thermostable form of soluble starch synthase in high-yielding thermosensitive cultivars of wheat. Research on the effects of brief periods of ear warming after anthesis on ear metabolism have identified differential responses of starch and nitrogen accumulation in the grains of four wheat cultivars (Bhullar and Jenner 1986; Hawker and Jenner 1993; Jenner 1991). Warming increased the rate of dry matter accumulation in all the cultivars, but the increase was less in the variety Aus 22645 compared to other cultivars. The rate of nitrogen accumulation was, however, higher than the increase in dry matter accumulation. Under long-term exposure to heat stress, the increase in grain nitrogen accumulation was attributed to the reduction in starch accumulation rather than a change in overall grain quality (Bhullar and Jenner 1985). The conversion of sucrose to starch within the endosperm is decreased by elevated temperature. Furthermore, heat stress effect on final grain weight was associated with reduced levels of soluble starch synthase activity (Hawker and Jenner 1993).

The net photosynthesis in wheat increases as temperature rises to a broad optimum before decreasing at high temperature (>25 °C) and ceasing at about 40 °C. The sensitivity of net photosynthesis (Pn) is predicted to increase with temperature due to characteristics of Rubisco particularly its kinetic parameters and the relative solubility of CO_2 and O_2 increasing the photorespiration with the rise in temperature

(Long 1991; Lawlor and Keys 1993). Higher temperature also accelerates leaf maturation and senescence reducing the duration of active photosynthesis. There is a variation in the response of respiration to the rise in temperature. The maintenance respiration was highly affected by higher temperature compared to growth respiration. High respiration rates especially during the night due to high temperature can increase reactive oxygen species leading to cell damage affecting pollen viability.

The development of wheat plants is the sigmoid function of temperature (Shaykewich 1995). It begins at a base temperature (3–5 °C) and accelerates slowly as the temperature rises and then faster as the temperature reaches about 30 °C and then slows rapidly. Mitchell et al. (1995) reported that a 1 °C warming would reduce crop duration by about 21 days (8 %), reproductive period by about 8 days (6 %), anthesis to maturity period by about 3 days, and grain-filling period by 2 days (5 %). However, the response may vary among the cultivars. Rawson (1988) suggested that high light intensity, ample water, and nutrients may have ameliorating effects on high-temperature stress. According to Russell and Wilson (1994), the extreme cold temperature may kill wheat plants and late frost induces sterility. Temperature below 5 °C and above 30 °C at anthesis adversely affects the pollen formation, thereby reducing grain set and decreasing the grain yield (Tashiro and Wardlaw 1990). Acevedo (1991a, b) suggested that breeding and selection of better adapted varieties may help in designing temperature-tolerant cultivars. White et al. (2011) observed that when wheat grains mature under very high temperature (>40 °C) especially with later planting dates, the grain-filling and developmental processes are interrupted causing premature senescence. Wheat spikes may respond more to heating in comparison to vegetative tissues. Although the duration of grain fill is largely determined by temperature, the lack of assimilate availability can accelerate grain maturation (Egli 2004). High temperature may reduce sink strength causing grains to mature earlier. These temperature responses may involve transcriptional control (Zhu et al. 2011) suggesting that plants may integrate temperature signals that are translocated to different organs (Fig. 4.4).

4.1.3 Floral Initiation and Development

Heat stress around floral initiation has severe effects on grain number. Grain number per spike decreased by 4 % for every 1 °C (from 15 to 22 °C) increase in the 30 days preceding anthesis (Fischer 1985). Castro et al. (2007) observed smaller grains in 14 wheat genotypes regardless of the duration and timing of high temperature. Increase in temperature (from 30 to 38 °C) during reproductive phase reduced the main shoot grain weight by 20–44 % (Tahir and Nakata 2005). Dias et al. (2008) observed that increase in day/night temperature from 25/14 °C to 31/20 °C can cause grain shrinkage through ultrastructural changes in the aleurone layer and endosperm cells. Yin et al. (2009) observed that 5 °C increases in temperature above 20 °C increased the rate of grain filling and reduced the grain filling duration by 12 days in wheat genotypes.

Temperature stress and grain development (Wheat)

*At low temperature

1. Grain development continued for longer time.
2. Higher grain weight

**At high temperature

1. Accelerate rate of grain growth
2. Greater respiratory loss
3. Faster senescence of the foliage and ears
4. Early termination of grain growth
5. Lower grain yield

Fig. 4.4 Temperature stress and grain development in wheat

1°C rise in global temperature:

Depresses grain yield by 8 to 10 %

Decline in grain number by 5.5 %

Reduction in grain weight by 3 %

Fig. 4.5 1 °C rise in global temperature

The biomass production decreases with increase in temperature due to the reduction in the duration of all developmental stages. The shortening of the maximum growth period will have more effect on biomass production than shortening of early growth stages. The negative effect of increased temperature on grain yield tends to be greater than the effect on biomass. The harvest index was significantly reduced under warmer temperature (Mitchell et al. 1995; Wheeler et al. 1996a, b; Batts et al. 1997). One degree Celsius increase in temperature during grain filling will shorten grain development period by 5 % and reduce harvest index and grain yield substantially. If high temperature at anthesis induces partial sterility, this also contributes to the greater reduction in grain yield (Mitchell et al. 1993) (Fig. 4.5).

4.1.4 Grain Quality

Wheat grain quality: Wheat-processing industries require specific grain quality parameters according to the demand of the product. The important quality parameters in wheat are grain hardiness, grain size, milling, dough strength, and protein and starch quality. These attributes are governed by genotype, interaction of climate, rising temperature, and fertility of the soil. About 95 % of the wheat produced is *T. aestivum*, which is utilized in making a variety of flat, leavened, and unleavened breads. The rest *T. durum* is used to produce semolina from which pasta is made. Grain protein content and alpha amylase activity are the grading factors in wheat trading. High alpha amylase activity adversely affects the rheological properties of the baking dough as it hydrolyses the excessive amount of starch. According to Abrol and Uprety (1969), Abrol et al. (1971), and Muller et al. (1998), the most important climatic factors which influence the rheological properties of gluten are temperature and fertilizer application. The environmental factor effect prior to anthesis is seen on seed germination, photosynthesis, tiller formation, and inflorescence development, influencing grain number (Herzog 1986). After anthesis, environmental conditions primarily affect the kernel size, protein content, and composition. Abrol and Uprety (1969) demonstrated that increased nitrogen fertilizer application improves the grain protein content and gliadin-glutelin ratio in gluten protein. Carbohydrate in the form of starch is also deposited in the grains, but its rates reaches the maximum at the grain-filling stage. When plants are exposed to the higher temperature after heading and during grain filling, the natural reaction of wheat will be to store less carbohydrate and more protein. The rate of protein synthesis in the grain is also promoted by warmth more than the rate of starch synthesis (Spiertz 1997). According to Randall and Moss (1990), the rise in temperature above 30 °C weakens the dough properties such as dough strength and loaf score due to increase in the proportion of gliadin protein. It was observed that high temperatures and low relative humidity were associated with reduced glutelin and increased gliadin contents, respectively (Blumenthal et al. 1993). The increase in temperature during grain-filling affects grain nitrogen concentration and reduces dough strength, gluten quality, and loaf rating. Generally grain nitrogen concentration increases with increase in temperature and reduced rainfall. Blumenthal et al. (1993) reported that in Australian variety trials extending over 27 years, grain nitrogen concentration was positively associated with number of hours above 35 °C during grain filling. He et al. (1990) in an experiment with durum wheat in China found that nitrogen content in grain was positively associated with the average temperature during grain filling and negatively associated with growth period. However, carbohydrate accumulation was adversely affected by rising temperature as the rate of senescence is increased, thereby reducing photosynthesis and grain growth. Studies on *T. aestivum* and *T. durum* genotypes sown at different planting dates in India demonstrated that high temperature reduced test weight and hectoliter weight and increased grain protein content. The impact was more pronounced in bread wheat than durum wheat cultivars. The increase in the protein content of bread wheat was 0.3 % per 1 °C, while in durum it was 0.15 % per 1 °C increase. The B-carotene in durum varieties remained between 4 and 5 umol mol^{-1} stable throughout the temperature range of 16–29 °C indicating its stability against temperature changes. Tester et al. (1995) indicated that warmer temperature substantially decreased mass per grain and starch content due to fewer and smaller type starch "A" granules and fewer amyloplast per endosperm. Starch gelatinization increased with the rise in temperature. The total and lipid-free amylase increased and amylolipid and lysophospholipids were not affected by high temperature. Such alterations may affect the nutritional quality and industrial use of wheat grains. Heat stress during grain filling affects grain protein content through the reduction in starch content. According to Daniel and Triboi (2000), the grain size is more affected by temperature than the quantity of grain nitrogen. Grain protein content was inversely related to grain size, while grain protein increases the functionality of protein decreases. The total protein yield of the area decreases because of reduction in yield under high temperature. Heat stress decreases the synthesis of glutenin, while the synthesis of gliadin remains stable and increases (Majoul et al. 2003). Heat stress also decreases sedimentation indices as an effect associated with increased level of essential amino acids (Dias et al. 2008). High temperature (37/17 °C) from flowering to grain maturity caused a significant reduction in the starch accumulation period in developing wheat grains compared to ambient conditions (24/17 °C, day/night). However, at extremely higher temperatures (37/28 °C), starch incorporation was completed 21 days earlier than ambient conditions, with an increased proportion of A-type starch granules (10–50 um diameter) and a decreased proportion of B-type starch granules (5–10 um diameter) (Barnabas et al. 2008). The elevated temperature brought about significant quantitative changes in oil composition in both starch and non-starch lipid fractions in wheat. The temperature-induced qualitative changes on fatty acid composition of wheat grain lipids such as increased conversion of oleate to linoleate markedly influence the milling and the baking properties of wheat flour (Fig. 4.6).

The high temperature in winter would hasten development and excessive tillering and make the crops subject to more frost damage. The adverse effect of frost damage could be amended by increased incidence of the rise in temperature around anthesis. The effect of annual climatic variation, increased temperature and CO_2 on crop growth rate (CGR), duration to anthesis, evapotranspiration, and grain yield was simulated for irrigated and rain-fed wheat crop using a crop growth model—WTGROWS. The effect of climate variations was much smaller in irrigated condition compared to rain-fed conditions. It also depends on the magnitude of temperature increase. A 1 °C increase in temperature reduced evapotranspiration and crop duration but had no effect on crop growth rate and grain yield under irrigated condition. However, such an increase in temperature under rain-fed condition causes a significant reduction in crop duration to anthesis but increased the crop growth rate (CGR) significantly. Further increase in temperature reduced grain yields of irrigated crops but had no significant effect on rain-fed crops (Aggarawal and Sinha 1993).

It is also suggested that the overexpression and under expression of individual endogenous genes and introduction of novel genes (disease resistance and grain composition) may give conducive results. The high-temperature-induced changes in basic metabolisms, such as in photosynthesis and growth, likely require more time. The future of wheat modeling in the field of global environmental changes increasingly emphasizes the ability to predict broad-scale sensitivity of wheat yield to the environment for diverse climates, soil, cultivars, and agronomy. This will entail to move to more top-down models with fewer parameters that are tested with large yield sets and remote sensing data, rather than single-site experiments.

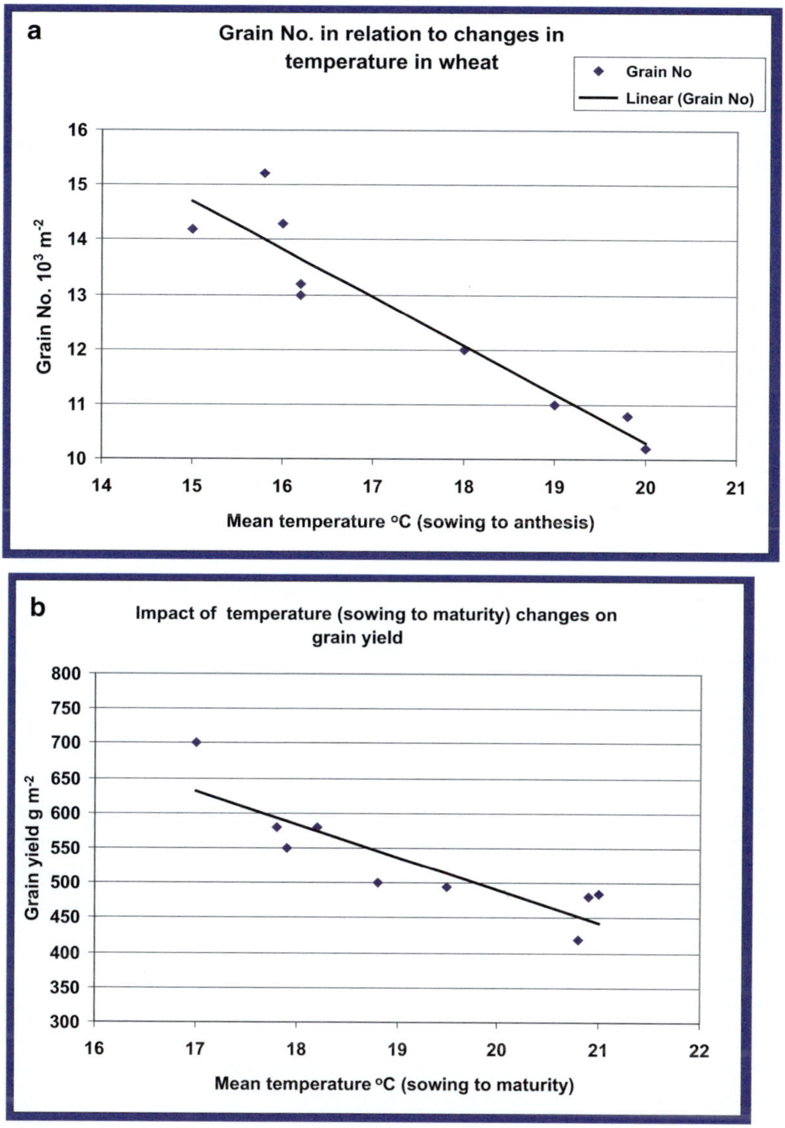

Fig. 4.6 (**a**) Grain No. in relation to changes in temperature in wheat (Abrol et al. 1992). (**b**) Impact of temperature (sowing to maturity) changes on grain yield

4.1.5 Mitigation

Tolerance Mechanisms of Terminal Heat Stress:
The adversities of heat stress can be minimized through agronomic strategies and by developing temperature-tolerant genotypes. Reduction of heat stress-induced damage in wheat plants is possible by leaf rolling, leaf shedding, reducing leaf

size, thickening of leaves, reducing growth duration, transpiration cooling, and other alternations in morphology and ontogeny (Wahid et al. 2007). Balla et al. (2009) demonstrated that heat stress during reproductive phase brings about an increase in antioxidant enzyme activities in heat-tolerant genotypes of wheat. The activity of catalase and superoxide dismutase has been correlated with heat stress (34/22 °C) during the reproductive phase (Zhao et al. 2007). Heat shock protein accumulates in the developing grains of heat-tolerant varieties more than susceptible varieties. Dehydrin protein of late embryogenesis abundant proteins (LEA) helps to stabilize the macromolecules against heat-induced damage. DHN-5 protein helped to protect and stabilize key enzymes for metabolic activities under high-temperature condition (Brini et al. 2010). Leaf senescence begins early during heat stress particularly when such stresses occur during post-flowering and grain-filling stages. Maintenance of leaf chlorophyll and photosynthetic capacity called "stay green" is considered as an indicator of heat tolerance. Stay-green genotypes are better able to maintain grain filling under elevated temperatures. Harris et al. (2007) reported that breeding for stay-green character has been possible in wheat. Three components of stay-green characters, chlorophyll content at anthesis, duration of senescence, and rate of senescence, determine the stay-green feature under heat stress. Membrane stability has also been found to be a suitable character for identifying a heat-tolerant variety. This character has been correlated with grain yield under high-temperature condition (Blum et al. 2001). The adverse effect of higher temperature might be overcome by selection of slower developing varieties, where water is not limiting. Various technological adjustments are required to mitigate the adverse effect of higher temperature in wheat production, which includes breeding by incorporating new genetic resources and by agronomic manipulations. Further research for genetic variability in the important traits, indicative of thermotolerance, is urgently required. It is highly important to use them in breeding programs to develop tolerant genotypes that can provide sustainable grain yield under current and future climates.

Agronomic strategies to counter the heat stress effect include practices that conserve water, fertilization during critical growth stages, and time of sowing. Continuous water supply to heat-stressed wheat helped to sustain grain-filling rate, duration, and size (Dupont et al. 2006). Application of NPK ameliorates the moderate heat stress effects in crop plants. The time of sowing is another important management strategy in some region. Early planting may avoid terminal heat stress so that grain filling occurs during cool temperature. According to Arnonson and Mc Nulty (2009), the supplemental irrigation is one of the important mitigation technologies to reduce the heat-induced vapor pressure gradients, and subsequently the secondary effect of the humidity and temperature is on stomatal conductance.

Seed priming with inorganic salts, plant water extracts, and organic molecules improves the tolerance to high temperature in crop cultivars. Priming treatment substantially improved the tissue water status, membrane stability, gas exchange, and crop productivity.

Adjustment in the time of sowing can minimize the effect of terminal heat stress in wheat. Timely sown wheat (in the first fortnight of November) escapes from heat stress, caused by a sudden rise in temperature in the last week of March. But terminal heat stress is common in late sown wheat in northwest India.

Investigations revealed that the adoption of cultural practices like seed priming with water and moringa water extract and ascorbate, timely sowing, mulching, additional light irrigation during post-anthesis, foliar spray of 0.5 % KNO_3 at 50 % flowering stage or 1 % KNO_3 during anthesis, foliar spray of 2.5 mM of arginine, and foliar spray of zinc can mitigate the adverse effect of terminal heat stress on wheat plants (Telfer et al. 2013; Sharma et al. 2003; Acevedo 1991a, b; Zhao et al. 2009; Sharma and Kumar 2010; Huang and Guo 2005; Mc Martin et al. 1974).

The total effect of global warming on wheat productivity cannot be predicted with confidence because of local changes in light intensity and evaporative demand and interactions with changing technology are not known. The impact assessment analysis of global warming will also require technological adjustments in wheat production by breeding, by incorporating new genetic resources, and by agronomic techniques.

4.2 Rice

Rice is the main food of the South Asian region. More than 85 % of the world rice is produced in Asia. Rice provides staple food for more than half of the current population and is predominantly produced in South, Southeast, and East Asia from Pakistan to Japan with China leading global production. Rice in terms of its food value provides over 21 % of the world's calorie needs (Fitzgerald et al. 2009). Recent global climate models predict an increase in mean temperature by 2–4.5 °C in rice growing area (IPCC 2007). Rice production is highly sensitive to increasing temperature (Peng et al. 2004). The air temperature range for its vegetative growth generally varies between 12 and 38 °C (optimum value ranges between 25 and 30 °C). The reproductive development in rice is affected by both low and high temperatures. Temperature below 20 °C inhibits pollen development and above 35 °C increases the spikelet sterility (Matsui et al. 1997). Rice crop cultivation in Southeast and East Asian region is more sensitive to low temperatures and less vulnerable to high temperatures. The cool temperature frequently results drastic reduction in the yield of rice in the northern parts of China, Korea, and Japan. The pollen sterility induced by the rise in temperature is generally recorded in tropical and subtropical rice growing areas during the dry season (Satake 1995).

The rise in temperature is projected to have negative effects mainly via:

1. A reduction in growth period
2. An increase in respiration rates
3. Depleted water resources due to larger water demand (Parry et al. 2004)

The rice production depends on the net balance between counteracting these adverse effects of high temperature and its interactions with other environmental and biological factors.

Current temperature is already approaching critical levels during susceptible stages of rice development in many Asian countries, and drought stress will be exacerbated as temperature rise (Wassmann et al. 2009). This crop is grown over variable climate and soil types including the dry areas between 45° N and 40°S. Most of the rice growing area falls in the regions where the temperatures are close to the optimal for the growth (28 °C/22 °C). Therefore, any increase in temperature, at the sensitive stages, will adversely affect the yield of rice cultivars.

4.2.1 Phenophase Stages

Seedlings of a rice begin as heterotrophic, depending totally on food mobilized from the endosperm, and then pass through a transition period when photosynthesis commences, while endosperm mobilization continues; finally seedlings depend entirely on photosynthesis and become autotrophic. Penning de Vries et al. (1989) estimated that 0.25 g/g^{-1} of seed dry mass was utilized for seedling growth under field conditions, whereas in some rice cultivars, this value is as high as 0.42 g/g^{-1} with seedling growth efficiency ranging 0.53–0.58 of transforming seed into seedling. They also demonstrated that about 14 % of dry mass produced between emergence and 50 % flowering was transported to roots and rest to shoots. The dry matter in the shoot was equally divided between culm and leaf during this period. Yoshida (1973) points out that temperature has a prominent effect on rice preautotrophic duration because the breakdown of seed reserve is enzymatic in nature. When temperature affects the duration of certain crop development phase, the use of thermal time is a better option than that of calendar day to express such duration. The growth of rice plants after transplanting accelerate linearly from 18 to 33 °C and adversely affected by high temperature in the tropical region and by lower temperature in the temperate regions. Rooting of rice seedlings occurs favorably over a range of 19–33 °C with an optimum of 25–28 °C (Chamura and Honma, 1973). Tiller production increases with the rise in the temperature in the range from 15 to 33 °C and the optimum temperature for tillering appeared to be 25 °C day and 20 °C night (Yoshida 1973). According to Chaudhary and Ghildiyal (1970), the temperature above 33 °C does not favor tillering. Panicle differentiation occurs in the range of 18–30 °C (Adachi and Inoue 1972). However, high temperature accelerates floral initiation (Vergara et al. 1972). It has been observed by Horie (1993) that high temperature at flowering and grain-filling period reduces yield due to spikelet sterility and reduction in grain-filling period. Temperature below 20 °C during the early microspore stage inhibits pollen development (Nishiyama 1984). The fertility of spikelet decreases with the increase in temperature above 30 °C due to desiccation of pollens. Even 1–2 h of higher temperature at anthesis results into a drastically high sterility (Satake and Yoshida 1978). Nakagawa et al. (2003) observed that heat-induced spikelet sterility occurs more frequently due to global warming and reported 16 % increase in spikelet sterility per 1 °C increase above the threshold temperature and almost all spikelets become sterile at temperature greater than 40 °C. Hasegawa et al. (2006) suggested that increased canopy temperature due to reduced transpiration is the main cause of spikelet sterility in rice. The projected global warming may decrease the incidence of chilling-induced sterility, which has been a major yield constraint under cool climates. Osada et al. (1973) demonstrated that the length of ripening period was adversely correlated with the rise in temperature, and, therefore, the grain filling is poor when the temperature is above optimum although the increase in temperature increases the rate of grain filling. Rice is most sensitive to low temperature at 2–1 week before flowering and at flowering. Cool night temperature below 15 °C for 10–14 days before heading causes high sterility of spikelet. Rice response to high temperature differs according to the developmental stage, with the highest sensitivity recorded at the reproductive stage. The temperature-induced greater spikelet sterility was attributed to abnormal anther dehiscence, impaired pollination, and poor pollen germination (Rang et al. 2011). Kim et al. (2011) observed that high temperature during grain-filling period resulted into lower grain weight and yield in rice cultivars. Their study showed that high temperature increased the rates of grain filling and leaf senescence while reducing their duration. Grain filling was terminated earlier than leaf senescence. This suggests that an early termination of grain filling under high temperature was not the result of the low assimilate supply due to the early senescence of panicle.

Hasegawa et al. (1999) demonstrated that suboptimal temperatures of irrigation water can affect rice growth by limiting many physiological activities at various stages of growth. The slow growth appears to be due to the reduction in root nutrient transportability caused by low irrigation water temperature. This also affects the reduction in tillering and leaf emergence. The reduction is attributed to the limited activity of leaf expansion that reduced the shoot demand. Indica varieties of rice are more susceptible to low temperature than Japonica varieties. Furthermore, Baker (2004) indicated that Southern US rice cultivars were more sensitive to high-temperature stress during reproductive development than Asian cultivars. These observations also point to the possibility of selecting and breeding rice cultivars with enhanced capability to take advantage of future global increases in CO_2 and temperature.

Yoshida (1978) listed critical temperature at various phenophases for the rice crop. Growth and development may be adversely affected if the minimum temperature goes below 12 °C. A few varieties at certain phenophase can tolerate temperature as high as 44–45 °C. Optimum temperature for cultivation varies from 18 to 33 °C. The temperature in the tropics usually remains within this limit. Traditionally cool temperatures have been more limiting for rice production than warm temperature. However, rice plants also respond to higher temperatures. High temperature promotes plant growth rate and results in reduced growth duration leading to shorter grain-filling period which varies from 25 days in the tropics to 35 days in the temperate zone (Swaminathan 1984). Spikelet sterility is induced by higher temperature (Yoshida and Parao 1976), which becomes very severe near 40 °C resulting a great loss in rice production. Peng et al. (2004) observed a reduction of 10 % in grain yield for each 1 °C rise in growing season minimum temperature in the dry season, whereas the effect of maximum temperature on crop yield was insignificant.

1 rice-transplanting

2 rice-flowering

3 rice-panicle initiation

4 rice-grain filling

4.2.2 Physiological Processes

According to IPCC (2013) report, the greater increase in mean night temperature (NT) and hotter maximum day temperature (DT) are detrimental to rice production and are important factors that need to be considered for future rice crop improvement. High night temperature influences the rice growth and production significantly. Although the high night temperature does not affect leaf photosynthesis, it has profound effects on chlorophyll content, leaf nitrogen, pollen germination, and spikelet fertility (Mohammad and Taropley 2009). Glaubitz et al. (2014), while studying the differential physiological responses of rice cultivars to high night temperature during vegetative growth, observed a significant increase in their respiration rates, whereas photosynthetic quantum yield was not affected. Monosaccharide and starch concentration in leaves were increased, while sucrose was not affected. The tolerant cultivars accumulated more monosaccharides without depletion of carbon from their leaves compared to vulnerable cultivars in which leaves are affected by chlorosis. High night temperature compared to high day temperature reduced the final grain weight by a reduction in grain growth rate in the early or middle stages of grain filling and also reduced the cell size midway between the central point and the surface of endosperm (Morita et al. 2005). High night temperature during reproductive growth stage reduced the stimulatory effect of elevated CO_2 on brown rice yield by decreasing the translocation of photosynthate to grains due to reduced spikelet fertility (Cheng et al. 2009).

In the wet season, the maximum air temperature is around 30 °C during the vegetative growth of rice plants; however, the higher temperature prevails at

5 rice-maturation

flowering and during grain-filling phase. Low temperature limits rice growth and yield at higher latitudes, i.e., sub-Himalayan rice growing areas during seedling and vegetative phase in the spring season and during grain-filling phase in the wet season. Growth rate increases linearly in the temperature range of 22–31 °C, and higher than this range may adversely affect rice growth and productivity (Yoshida 1981). Rice plant requires about 2000–4000°days which corresponds to 80–160 days when grown at a mean temperature of 25 °C. According to Yoshida (1981) and Venkataraman (1987), the optimum temperatures for the following physiological processes and growth stages are recorded:

Physiological processes	Temperature (°C)
1. Cell division	25
2. Cell elongation	30
3. Photosynthesis	20–35
4. Respiration	20–32
5. Translocation	18–29
6. Germination	20–35
7. Seedling emergence and growth	25–30
8. Rooting	25–28
9. Leaf and stem elongation	22–31
10. Tillering	25–31
11. Panicle initiation	25–30
12. Anthesis	30
13. Ripening	20–25

The optimal temperature for ripening is lower than that for tillering and anthesis because temperature optimum shifts to relatively lower temperature as rice grows. Considerable reduction in the production of rice caused by high temperature would exceed further under future global warming scenarios. However, relatively large variability exists in the response of rice cultivars for tolerance to high and low temperature which may help in designing and developing plant types for the future warm world. Wang et al. (2014) indicated that temperature will remain a principal factor for determining the rice productivity for the future. The climate warming during 2021–2050 would substantially reduce the cold stress and increase heat stress on the productivity of rice crop. The normal growing degree days (NGDD) and killing growing degree days (KGDD) will remain important factors in determining the rice grain yield. The revised models will be required to determine and predict the future rice production (Fig. 4.7).

A close linkage between rice grain yield and mean minimum temperature was observed during dry cropping season (January–April). Observations of Peng and his group (2004) supported this concept showing 10 % reduction in rice grain yield for each 1 °C rise in minimum temperature under dry conditions. Huang et al. (2013a, b) in a study at low-temperature condition observed 7 % increase in rice grain yield for each 1 °C increase in growing season temperature. This was

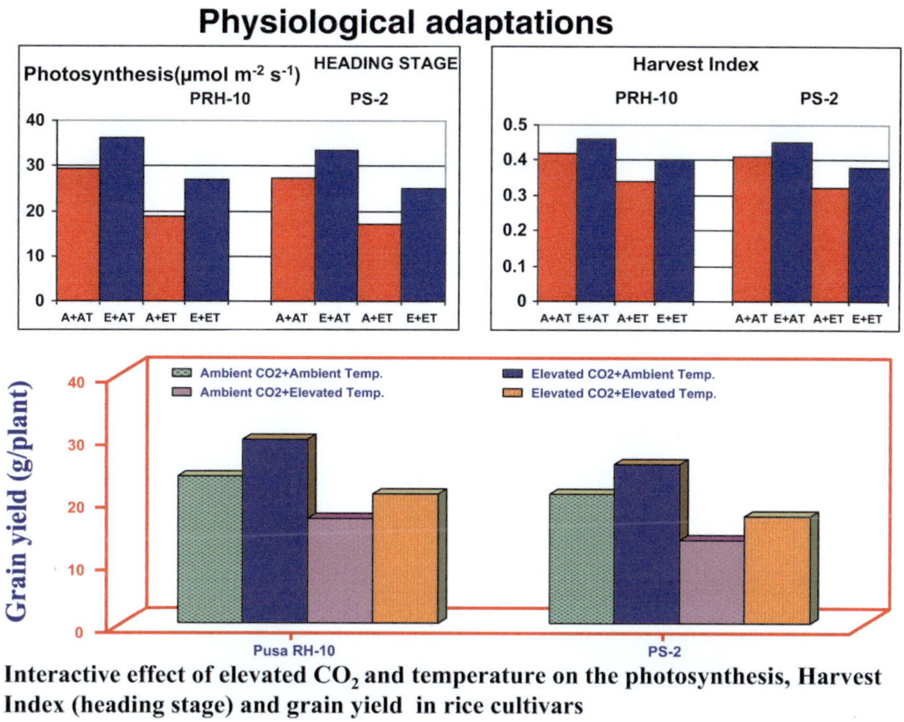

Fig. 4.7 Physiological adaptations of rice under elevated CO_2 and temperature (Uprety and Reddy 2008)

attributed to increase in per day yield rather than the duration of growth. They have recorded that transplantation could sustain the productivity of rice cultivars under intermittent changes in the growing season temperature. The direct seeding is a high-risk establishment method under changing temperatures for the early rice production.

4.2.3 Elevated CO_2 and Temperature Interaction

The rise in both temperature and CO_2 influences crop growth; therefore, the understanding of crop responses to the interaction of elevated CO_2 and temperature would be highly significant. A study of interactive effect of CO_2 enrichment and temperature on carbohydrate accumulation and partitioning showed that optimum temperature for total nonstructural carbohydrate accumulation was 25 °C for ambient and elevated CO_2 conditions, but plants grown in CO_2-enriched conditions could survive and maintain carbohydrate production rates at higher temperature than plants grown under ambient CO_2 level (Rowland-Bamford et al. 1996). Positive $CO_2 \times$ temperature interaction on leaf photosynthesis of rice plants was observed

Interactive effect of elevated CO_2 and temperature on grain quality parameters in the rice varieties

Treatments	1000 Grain Mass(g)		TNC (mg g^{-1} DW)		Protein (%)	
	PRH-10	PS-2	PRH-10	PS-2	PRH-10	PS-2
Amb CO_2 + amb T.	26.8	24.9	523.6	505.7	9.5	9.4
Elev CO_2 + amb T.	28.1	26.9	559.5	539.9	9.4	9.2
Amb CO_2 + elev T.	18.3	16.9	373.7	357.9	9.1	8.8
Elev CO_2 + elev T.	23.6	19.9	434.4	417.6	9.2	8.9
	Total sugar (mgg^{-1} DW)		Starch (mg g^{-1} DW)		Amylose (%)	
	PRH-10	PS-2	PRH-10	PS-2	PRH- 10	PS-2
Amb CO_2 + amb T.	250.3	245.3	258.3	256.9	24.3	26.1
Elev CO_2 + amb T.	270.1	262.3	289.4	277.6	25.0	27.5
Amb CO_2 + elev T.	179.4	171.7	194.2	175.2	23.9	25.0
Elev CO_2 + elev T.	208.1	197.3	226.3	198.3	24.0	27.1

Fig. 4.8 Interactive effect of elevated CO_2 and temperature on grain quality parameters in the rice varieties (Uprety 2007)

within limits by Imai and Okamota-Sato (1991), but Baker and Allen (1993a, b) reported relatively insensitive interaction under dense population of crop communities. Canopy photosynthesis in general is relatively insensitive to a wider range of surface temperatures (Jones et al. 1985). CO_2 enrichment causes partial stomatal closure, increased stomatal resistance, reduced leaf and whole canopy transpiration, and warmer leaf and canopy temperature (Baker and Allen 1993a, b). The single leaf photosynthesis in rice is reported to be stimulated at higher temperatures during vegetative stage. At the same time, the canopy photosynthesis was found to be unaffected by the changes in temperature. The lowered canopy surface temperature due to greater transpirational cooling at higher air temperature may be one of the reasons for this different photosynthetic response to temperature between single leaf and whole canopies (Baker and Allen 1993a, b) (Fig. 4.8).

At flowering stage, photosynthetic stimulation by elevated CO_2 appeared to be accompanied by a reduction in Rubp carboxylase/oxygenase activity as evidenced by the reduction in the assimilation at a standard internal CO_2. High temperature can reduce the photosynthetic rate by 40–60 % at mid ripening leading to more rapid senescence of the flag leaf (Oh et al. 2007). High temperature also results to the demand of a greater sink size due to increased respiration and plant growth resulting into the rapid use of assimilates. Vu et al. (1997) observed that CO_2 exchange rate (CER) in rice leaves increased by CO_2 enrichment; however, it decreased by the rise in temperature. The rise in temperature adversely affects the kinetics of Rubisco enzyme (Gesh et al. 2003). Higher temperature affects the larger subunit of Rubisco (RLSU) more than small subunit (RSSU) in tolerant rice cultivar like N22, but no such trend was recorded in temperature-sensitive paddy species IR8. The tolerant cultivar showed greater thermostability of Rubisco protein. The photorespiration will increase with temperature largely because of the reduction in the specificity of Rubisco for CO_2 and its activation. Nagai and Makino (2009) recorded a greater reduction in the ratio of chlorophyll a/b in rice leaves of plants grown in high temperature. The elevated temperature brings about an increase in the rate of photosynthesis at tillering stage; however, at flowering the photosynthetic activity was adversely affected under high temperature due to reduction in the Rubisco activity. The high temperature reduced the net photosynthetic rate (Pn), the apparent quantum yield (AQY), the photochemical efficiency of PS II (Fv/Fm), the quantum yield of PS II electron transport (Q PS II), and the coefficient of photochemical quenching (qp) but increases the relative reduction state of PS II (Guo and Lee 2006). The higher-temperature-induced increase in respiration may utilize a greater proportion of photoassimilate. The tissues and organs have less time to acquire photoassimilates resulting into reduction in the size and number of organelles and reduction in biomass production (Zheng et al. 2012).

Ziska et al. (2014) in an exclusive study on the responses of cultivated and wild weedy rice crop lines demonstrated that the relative increase in seed yield with CO_2 and air temperature was positively associated with panicle and tiller number but negatively associated with the percentage of immature seeds. This study identified that variety "Stg-S," a weedy red rice species with high tillering capacity, was able to demonstrate a significant increase in seed yield under high CO_2 and elevated atmospheric temperature. The early differences in tiller formation may be an effective means to facilitate screening for high-temperature and high CO_2-responsive rice cultivars.

Ingram et al. (1995) reported significantly greater tiller production under higher temperature (37/29 °C) and elevated CO_2. However, higher temperature shortened the tillering duration. Relatively lower temperature (33/25 °C) produced tillers for a longer duration. High night temperature and high CO_2 result into late tillering (Imai et al. 1985). Rice cultivars vary in their tolerance to high-temperature stress (37 °C) during flowering. The ability of elevated CO_2 to mitigate high-temperature effect on filled spikelet percentage depends on the susceptibility of the cultivars to high temperature under ambient CO_2 (Ingram et al. 1995). Ziska and Bunce (1997) reported that larger yield reduction caused by high temperature occurred at elevated CO_2 in the dry season. Matsui et al. (1997) and Kim et al. (1996) recorded a negative interaction between CO_2 and temperature on seed set and yield. Hamid et al. (2003) reported that elevated CO_2 compensated the negative effect of higher temperature on sterility and chlorophyll ratio sustaining the productivity of rice cultivars. Sujatha et al. (2008) observed that the sucrose-P synthase (SPS) activity and the accumulation of soluble sugars and starch were enhanced at elevated CO_2 under normal and elevated temperature regimes. They concluded that the upregulation of SPS under elevated CO_2 and temperature would be beneficial for growth and productivity of rice cultivars for the future climatic conditions.

The increasing concentration of CO_2 in the atmosphere enhances the rate of phenological development and shortens the time duration to heading. This reduction has been considered as temperature dependent. The reduction in a number of days to heading was 11 % due to temperature rise and 6 % due to CO_2 enrichment in *Japonica*-type rice (Kim et al. 1996). This was attributed to the elevated CO_2-induced higher plant temperature by increasing stomatal resistance and reducing transpirational cooling. This can also be explained by higher C-N ratio, which accelerates the plant development rates. Horie et al. (2000) reported 30 % increase in rice grain yield under doubling CO_2 concentration, but an enhancement in temperature caused a reduction in yield. Kim et al. (1996) demonstrated that the rice grain yield is more sensitive to the rise in temperature under high CO_2 than that of ambient CO_2. It was due to the shortening of grain-filling period and lower spikelet fertility in high CO_2-grown plants.

More experimental studies are required to quantify the effects and interactions of CO_2 and temperature on biomass production, stomatal conductance, water use efficiency, spikelet sterility, and grain yield. Studies are also needed to understand the mechanisms controlling the effect of high temperature-induced spikelet sterility by an elevated concentration of CO_2. Variability among rice genotypes in response to CO_2 and temperature in the areas of temperature-induced spikelet sterility, water use efficiency, and productivity needs to be studied. This will help in identifying rice genotypes that are better adapted for future climate change. Models are needed for simulating climate change effects on rice to predict the effect of rising temperature, CO_2, and other environmental factors on growth and yield formation processes and to develop strategies for sustainable rice production in changing the climate.

4.2.4 Physiological Processes

The dark respiration might be one of the potential causes of high crop growth rate in the regions, where the day/night temperature difference is large (Lee and Akita 1999). Yamada et al. (1955) observed that temperature coefficient (Q10) in rice leaf was 1.66–2.06 during the whole growth stage and it was highest at booting stage. The dark respiration of rice shows a circadian rhythm. It decreases after sunset and begins to increase after midnight till morning in continuous darkness. The rhythmic behavior of dark respiration is attributed to the sequential change in the amount of respiratory substrates. The marked changes in dark respiration could be one of the factors responsible for the high growth rate at the ripening stage.

The optimum temperature for phenological development of rice toward heading is generally 27–32 °C (Yin et al. 1996), and in most cases, Indica-type rice has a 1–3 °C higher optimum temperature than Japonica-type rice. The reproductive processes that occur within one hour after anthesis-dehiscence of anther, shedding of pollen, germination of pollen grains on stigma, and elongation of pollen tubes are more sensitive to higher temperature and are disrupted at day temperatures above 33 °C (Satake and Yoshida 1978). Similarly night temperature more than 29 °C increases the susceptibility of rice to sterility with a subsequent reduction in seed set and grain yield (Ziska et al. 1996). Thus, the main effects of high-temperature stress during flowering were:

1. Decreased pollen production
2. Indehiscence of anthers prematurely causing poor pollen shed
3. Decreased number of pollen grains intercepted by the stigma. According to Prasad et al. (2006), the differential response of high temperature was attributed to the differences in spikelet fertility and harvest index among rice cultivars. The number of spikelets per unit biomass at the heading stage decreased with increasing temperature above 24 °C irrespective of the other climate change parameters. The spikelet sterility in rice is due to the failure in pollination at mean daily temperature below 20 °C at the young microspore stage of pollen development. The adverse effect of high temperature for grain filling in rice was initiated by the flowering stage temperature above 35 °C (Satake and Hayashi 1970). Through these three processes in rice response to temperature, rice yield generally declines with increased temperatures above 25 °C. The Japonica rice grown under doubled CO_2 showed more severe yield reduction with the rise in temperature than that under ambient CO_2. This decline in yield was associated with significantly lower number of fertile spikelets at higher temperature. The reduction in the fertility percentage was mainly due to high temperature-induced spikelet sterility during flowering (Matsui et al. 1997). They also demonstrated that daily maximum temperature was more closely related to the damage to spikelets than the daily mean temperature.

Baker and Allen (1993a, b) demonstrated a reduction of 10 % in rice grain yield per 1 °C rise in temperature above 26 °C. The decrease in grain yield was attributed mainly to the fewer grains per panicle. High day and night temperature considerably reduces the viability of pollen grains at anthesis, increases the floret sterility, and decreases the seed set in rice. This heat-induced increase in floret sterility may have been responsible for the downregulation of photosynthesis observed in rice under high temperature (Lin et al. 1997). Yoshimoto et al. (2005) in a FACE experiment observed that high CO_2 induced 12 % reduction in stomatal conductance, decreases water use, increases water use efficiency, but at the same time increases the canopy and panicle temperature about 1 °C higher than ambient condition exacerbating the heat-induced spikelet sterility. Whereas the natural temperature rise due to global warming may decrease the incidences of chilling-induced sterility, which has been the major yield constraint under cool climate.

Ashikari et al. (2005) found a gene to increase spikelet number per panicle through the accumulation of cytokinin in inflorescence meristems. The morphological characteristics of panicle in rice genotypes with larger spikelet numbers are probably regulated by cytokinin dynamics. Horie et al. (2007) showed the degeneration of spikelets increased exponentially with decreasing crop growth rate (CGR) during the late reproductive period. It showed that when CGR is as low as 10 gm^{-2} day^{-1}, about 40 % of differentiated spikelets degenerate. Horie et al. (2003)

showed that the genotypic differences in CGR during the late reproductive period were mainly due to the difference in leaf photosynthetic rate. The photosynthate produced during the initial grain-filling period is not enough to supply the necessary amount of carbohydrates for all the spikelets in a panicle for their full development. Thus, the nonstructural carbohydrates stored in the plants during the reproductive period play a key role in grain-filling process.

4.2.5 Rice Models

Crop models have been widely used in simulating and predicting changes in rice phenology in the rice-producing areas of China (Zhang and Tao 2013). These model predictions were generally consistent when the temperature was below optimum and the simulation was different when the temperature exceeded above optimum. IRRI crop modeler Sheehy et al. (2006) determined that for every 75 $umol^{-1}$ increase in CO_2 concentration, rice yields will increase by 0.5 tonne per hectare, but yield will decrease by 0.6 tonne per hectare for every 1 °C rise in temperature. Sheehy et al. (2006) used the growth models ORYZA 2000 and EEQ to estimate the effect of temperature change within −2 to +2 °C on the change in yields and observed yield decrease of 0.5 t/ha/ 1 °C or 6 %/1 °C from a base yield at 26 °C. Lal et al. (1998) used CERES rice model and predicted a 20 % decline in rice yields in northwestern India due to elevated CO_2 and temperature. Pasquin et al. (2008a, b) showed that higher grain yield was related to higher shoot dry weight, but at much higher temperature, grain yield was related to number of filled grains per panicle. Therefore, the rice responses to elevated temperature are highly dynamic in terms of source sink relationships. Thus, there is a need to expression based on temperature responses of these components over growth stages. Krishnan et al. (2007) predicted average yield changes of −7.2 % and −6.66 %, respectively, for every 1.8 °C increases in temperature using ORYZA 1 and InfoCrop rice simulation models. Studies made under doubling CO_2 resulted in an average yield increases of 30.73 % by ORYZA 1 and 56.37 % by InfoCrop rice. At a temperature increase +4.8 °C above ambient temperature, the differences between two models become small. Further, GDFL, GISS, and UKMO scenarios, ORYZA 1 predicted the yield changes of −7.63, −9.38, and −15.86 %, respectively, while InfoCrop predicted changes of −9.02, −11.30, and −21.35 %. There were considerable differences in the yield predictions for individual locations due to higher-temperature-induced sterility of rice spikelets; however, the trend was similar.

4.2.6 Pests and Their Management

The combined effect of increased temperature, elevated UV-B radiations, and rice blast disease (*Pyricularia grisea*) on rice yield was assessed using a coupled simulation model (Luo et al. 1997). Kaukoranta (1996) developed degree day models for the emergence of potatoes and the date of late blight outbreaks. The two models were coupled and validated and used for various temperature change scenarios to predict possible changes in yield and yield losses in crops caused by late blight and blast diseases in a warmer climate. The insect population dynamic model for gundhi bug (*Leptocorisa acuta*) was being updated daily depending upon the rate of population change, total insect numbers in each development stage. Chander et al. (2009) concluded that up to 1 °C rise in daily average temperature over current temperature would not affect gundhi bug population much, but the further increase would cause a significant decline in it.

4.2.7 Grain Quality

Rice is an important food crop, which provides 21 % calorific need of the world population. Quality traits like physical appearance, cooking and sensory properties, and nutritional properties define the market value of this crop. Physical properties include shape of the grain, whiteness and translucence of grain, and the cooking qualities which include cooking time, textural properties of cooked rice, aroma and its sustenance after cooking, and retention of the softness of cooked rice for a longer period. Grains with opaque areas in the endosperm (chalkiness), caused by the loose packing of starch and protein particles, break more readily than clear grains during milling, greatly reducing the market value. Amylose content in the grains determines the milling and cooking properties of rice. Japonica varieties have low amylose content and are sticky on cooking. Indica varieties vary widely in amylose content according to regional quality preferences. These grain quality parameters of rice are strongly influenced by higher temperature prevailing during grain development phase, and its impact varies with cultivar and its adaptability to the climate change.

Huang et al. (2013a, b) in a study on early season rice cultivar between two contrary years with respect to temperature, i.e., 2009 (normal temperature) and 2010 (low temperature) at seedling stage, demonstrated that rice grains at low-temperature year showed lower milled rice percentage, reduction in gel consistency, and gelatinization temperature. The low-temperature year also showed a significantly high percentage of chalky rice grains, degree of chalkiness, and protein content. The shortened grain-filling duration indirectly caused by low temperature at seedling stage was responsible for decreased grain weight and poor quality. Coast et al. (2015) demonstrated that higher night temperature (HNT) reduced spike fertility, grain weight per panicle, amylose content in grains, and gel consistency, whereas day temperature reduced only gel consistency. They observed that high night temperature was more damaging to grain weight and grain quality traits compared to high day temperature.

Warmer temperature (more than 22 °C) interferes with metabolic processes during developmental stages in rice affecting yield as well as pre- and postharvest quality. Chalkiness, a major quality defect in rice kernels, occurs due to increased temperature during critical developmental stages of growth. Severely chalky brown rice grains are inferior for policing quality and palatability. The microscopic observations of the chalky parts of the high-temperature-ripened grains revealed that loosely packed starch granules create air spaces between themselves to reflect light randomly (Zakaria et al. 2002). This chalkiness was attributed to the irregular packing of starch granules (amyloplasts) which occurs during grain-filling stage (Ashida et al. 2009). Warmer temperature induces the chalkiness due to decreased supply of substrates, which not only affects the grain-filling processes but also decreases the expression of genes that control enzyme activity during starch synthesis (Sujatha et al. 2008). These chalky kernels are weaker and prone to breakage during milling. The high-temperature stress during middle to late stages of grain-filling results in the adverse effect on the grain quality of rice. Krishnan and Rao (2005) observed that while the yield attributes were influenced by genotypic effects, the seed quality traits such as seed leachate conductivity, potential seed longevity, percentage seed germination, and the seed discoloration were influenced by environmental factors. Warm weather (temperature more than 28 °C) without excessive rains during grain filling produces best rice yield with superior grain quality. Aromatic rice requires relatively cooler temperatures (less than 25 °C day and less than 21 °C night) during crop maturity for better retention of aroma. Ghosh et al. (2004) found that delayed planting recorded lower milling as well as poor head recovery and reduced amylose content. High day temperature during grain ripening period of the late-planted crop resulted in a number of weak kernels having sun cracks that lowered head rice yield. According to Rahman (2003), the detrimental effect of low temperature due to delayed sowing was more pronounced in photosensitive varieties. Mohapatra et al. (1993) observed that the spikelets anthesed earlier produced better quality grains with higher sink efficiency in converting assimilates into reserve material than those which anthesed later in tall Indica rice cultivars. According to Venkatswarulu et al. (1987), the proportion of high density grains linearly increased with increase in photosynthetically active radiation after anthesis. Nagrajan et al. (2010) in a study on aromatic (Pusa Sugandh 2, Pusa 1121, and Super Basmati) and nonaromatic (Pusa-44 and IR-64) rice observed that the slenderness of grains was reduced in all the cultivars except Pusa-44, if the mean temperature during grain growth was either higher than 28 °C or lower than 24 °C. Among the aromatic varieties, Pusa Sugandh was less sensitive to temperature changes than other genotypes. Similarly, Khush et al. (1979) reported maximum elongation in Basmati rice at 25/21 °C day/night temperature during grain ripening and explained the greater elongation of Basmati rice, grown in Punjab than that grown in Sindh. Temperature above 25 °C adversely affected the thousand grain weight to the extent of 2–3 g per 1 °C rise in temperature. Amylose content of the grain decreased linearly with increase in mean temperature. This reduction varied from 0.65 % per 1 °C rise in temperature in IR-64 to 0.30 % in Super Basmati. Higher temperature after flowering raises the gelatinizing temperature (which lowers the grain quality). This character increased at the temperature above 30 °C in Pusa Sugandh 2 and IR-64 and was stable in other cultivars. Basmati rice requires relatively cooler temperature (25/21 °C day/night temperature during crop maturity) for better retention of aroma (Mann 1987). Aroma in Basmati cultivars P-1121 and Super Basmati declined drastically, when the mean temperature during grain development increased above 25 °C and in Pusa Sugandh 2 when the temperature was above 29 °C (Nagrajan et al. 2010).

Yamakawa et al. (2007) in a study on grain-filling-related genes under high temperature revealed the downregulation of several genes for starch or storage protein synthesis such as GBSSl, BEllb, and prolamin and upregulation of genes for starch consuming amylase genes were identified. Expression of several redox reaction-related genes was changed by high temperature in the microarray analysis. Asatsuma et al. (2006) reported ectopic overexpression of amylase genes resulted in grains with decreased weight and chalky appearance even ripened under ambient temperature, and antisense suppression of sucrose transporter gene (SUTl) decreased grain weight in transgenic rice plant (Scofield et al. 2002). Expression of amylase genes and sucrose transporter genes upregulated and downregulated by high temperature during ripening, respectively, suggests that these alterations in gene expression possibly contribute to grain weight reduction and/or grain chalkiness.

Lin et al. (2009) reported that higher temperature decreases the amylose and total starch content in the grains of rice cultivars. It increases the value of maximum viscosity and breakdown value but decreased the value of setback for Japonica variety. Conversely, the starch viscosity characteristics of Indica varieties were unaffected by elevated temperature. The high air temperature increased the absolute value of grain protein in Indica (IR 72) but not in Koshihikari (Japonica rice). However, the absolute content of some amino acids did not increase due to the reduction in grain weight under elevated temperature. Higher temperature also brought about the reduction in rice milling percentage and the appearance of grains; however, the cooking and eating qualities of the Japonica rice cultivar were not affected.

4.2.8 Mitigation

1. Identifying and developing heat-tolerant cultivars will be a vital task for rice breeders to meet the food requirements of growing populations in future climates. Spikelet fertility could be adopted as a screening tool for identifying high-temperature-tolerant cultivars. Zhang et al. (2013) identified cultivars, namely, IR 8, Swarna, N22, and PSB, as insensitive to high night temperature, whereas cultivars like Sambha Mahsuri, IR6, BRRI Dhan 29, and Saunfi were found to be the most sensitive to higher night temperature. Coast et al. (2015) in an exclusive study observed a novel high-temperature tolerance complementary mechanism of the shorter flower open duration in the rice variety N 22. They have identified the critical tolerance and escape traits such as spikelet cooling, higher spikelet fertility, early start, and peak time of anthesis in the cultivar N 22. Such cultivars can help in rice breeding program for rising temperature.
2. The shifting time of day of anthesis to early hours of the morning will help plant to escape high-temperature stress during the process of pollen shed, pollination, and fertilization. This can minimize the sterility caused by high temperature.

There is a potential for genetic improvement to advance flowering to an earlier time of day in high-yielding cultivars (Nishiyama and Blanco 1980). Growth regulator such as MeJA helps in advancing the time of flowering to early morning to avoid the effect of heat stress.

3. The use of biotechnology for developing temperature stress-tolerant varieties is one of the modern methods. A chromosome 3 QTL has a large effect on the drought and temperature tolerance in the cross between a tolerant variety APO and widely grown susceptible variety Swarna (Venuprasad et al. 2009). It is a promising QTL for use in marker-assisted selection because var. Swarna is widely grown in drought-prone area due to its high yield and other desirable traits.
4. Exogenous application of growth regulators such as glycine betaine (GB) and salicylic acid (SA) has some positive effect on spikelet fertility and pollination by increasing the level of endogenous antioxidants and preventing the oxidative damage to the membrane in rice (Mohammed and Tarploy 2009, 2011).
5. Flowering at cooler times of the day, more pollen viability, large anthers, longer basal dehiscence, and presence of long basal pores are some phenotypic markers for high-temperature tolerance.
6. Protection of structural proteins, enzymes, and membranes and expression of HSPs are some of the biochemical processes that can impart thermotolerance in rice cultivars. High temperature induces the expression of HSPs and suppresses partially the synthesis of normal cellular proteins.
7. Site-specific adjustment in cropping systems may be needed because the effect of these changes in temperature varies with the region.
8. Management practices such as saturated soil culture (SSC), alternate wetting and drying (AWD), and aerobic rice cultivation in the cases of drought stress while growing improved varieties of rice containing the sub 1 gene in flooded soils offer some adaptive options for higher-temperature stress.

All these traits will help in converting heat-sensitive cultivars into heat-tolerant cultivars. Adjustment of sowing time, choice of varieties with a growth duration allowing avoidance of peak stress periods, and exogenous application of plant hormones are some of the measures that will help in the mitigation of yield reduction due to global warming.

Moderately large genetic variation in the tolerance to high-temperature-induced spikelet sterility has been reported among Indica- and Japonica-type rice genotypes. Some rice cultivars have the ability to flower early in the morning, thus potentially avoiding the damaging effects of higher temperature later in the day (Imaki et al. 1987). The continued search for rice genotypes having a higher tolerance to and/or the ability to avoid high-temperature damage to spikelets is important in terms of finding options for mitigating potentially adverse effects of rising temperature on rice production.

According to Mackill et al. (2010), progress on developing rice varieties for the unfavorable climatic conditions has shown that modern breeding tools can address many of the problems of farmers in these areas. Furthermore with improved crop and soil management practices, these farmers can improve the rice yield under highly stressed condition.

Strong efforts are required to minimize the impact of warming on rice productivity and to sustain the production on the changing environment. Strong science will decipher this by using improved germplasm adaptations to expected future climatic conditions and mitigate the negative effect of warming on rice productivity. It is also expected that genomic sciences and comparative biology may be able to break the yield ceiling of rice and enhance its water and nitrogen use efficiency by changing the photosynthetic mechanism in rice to that of more efficient C4 plants. Efforts are being made to develop a C4 rice plant. A shift toward C4 system can potentially increase C uptake, but it may also lead to greater CO_2 losses from heterotrophic soil C respiration.

The effects of global warming on rice production will not only negatively affect farmer's income but also influence adversely the poor populations due to its secondary effect on price rise. The impact of this severe effect of warming on rice production would directly influence the food calorie availability and child malnutrition. Farmers can adapt to climate change by shifting planting dates, selecting cultivars with different growth duration, or changing crop rotations. Seasonal weather forecasts could be one supportive measure to optimize planting dates. Conservation and improved management of water contribute significantly in mitigating the heat stress in rice crop. Resource-conserving technologies such as enhancing input use efficiency provide economic benefits such as reductions in production costs, saving in water, fuel and labor requirements, and timely establishment of crops resulting in improved yields. These resource-conserving technologies, such as zero tillage, laser-aided land leveling, direct drill seeding of rice, diversification, raised bed planting, and leaf color chart for nitrogen management, may bring sizable economic benefit to farmers. The farmers of rice wheat belt have taken initiative to diversify their agriculture by including short-duration crops such as potato, soybean, mung bean, cowpea, pea, mustard, and maize in different combinations (Pandey and Sharma 1996). With appropriate crop establishment technologies, it is possible to raise a second crop of wheat, pulses, maize, or lentils to improve the income of farmers and their livelihood in marginal areas.

The rice cropping system is the economic backbone of most of the South Asian region. A small decrease in rice productivity will drastically imperil food security of South Asian countries. Therefore, this cropping system needs to be modified or diversified to increase its resilience to the rise in temperature. One approach is to improve germplasm to provide higher tolerance to temperature stresses (Ortiz et al. 2008) by selecting cultivars that exhibit heat tolerance during reproductive development, high harvest index, small leaves, and low leaf area per unit of growing area. Heat avoidance mechanism such as efficient transpirational cooling is found in germplasm of some arid regions. Another strategy to overcome high temperature is to shift the daytime of peak flowering to cooler periods. Research is also in progress for reconfiguring the rice photosynthetic system to C4 system to make rice plants less vulnerable to the rise in temperature.

Changes in temperature will affect pest and disease incidences. Some adaptation strategies, such as developing varieties resistant to pest and diseases, adoption of integrated pest management, and pest and disease forecasting using simulation models as tools, can reduce the infestation damages. Forecasting and early warning systems will be useful to minimize the risk of climate stresses. Crop insurance schemes needs to be incorporated in government policies to help farmers in reducing the risk of crop failure due to these events. The integration of local knowledge of climate change adaptation needs to be harnessed as more culturally appropriate options. Farmers use indigenous knowledge to minimize the losses by making changes in sowing and planting times, methods of sowing, variety, harvest time, fertilizer use, and seed rate.

It is proposed to identify temperature vulnerable regions so that the tolerant rice cultivars could be tested and accordingly the crop management program for them could be developed. This will be crucial for planning targeted adaptation programs to sustain the optimum production. There is a need to develop a policy framework for implementing the policy options so that farmers are saved from adverse impacts of climate change. The scientific progress needed in understanding the physiology of abiotic stresses and in developing biotechnology tools to mitigate the impact of climate change stress in rice crop productivity. More research programs on climate change adaptations are required to prevent drastic yield losses in this important crop that would bring social stability and economic development in the whole South Asian zone.

4.3 Soybean

Soybean is one of the 30 important food crops of the world. It is an important legume grown for its oil and protein. Its area of cultivation is more than 74.4 million hectare in the world and about 29.3 million hectare in the USA (Salem et al. 2007). Although the dry matter production in soybean is lesser than many important cereal crops, it is important because of its higher protein and oils, which contribute to human and animal diets. Soybean requires 100–150 days for the completion of its life cycle, i.e., from sowing to maturity. Soybean can be produced over a mean daily temperature range of 20–30 °C. However, the night temperature below 12 °C and a higher day temperature more than 36 °C adversely affect its production. Sionit et al. (1987) found that the rate of leaf appearance, leaf expansion, and branching increased by high temperature up to a range of 26/20 °C day/night temperature, respectively. The rates of soybean leaf appearance, leaf expansion, and branching increased up to 30 °C (Hofstra and Hasketh 1975). Rates of soybean leaf appearance and leaf expansion increase up to 30 °C, relative growth rate increases up to 31 °C, leaf photosynthesis rate increases up to 35 °C, and total biomass increases up to 28 °C (Allen and Boote 2000). Sionit et al. (1987) demonstrated that increasing temperature and increasing CO_2 concentration decrease the plastochron interval. Their study also showed that the temperature effect on soybean developmental rate was greater than that of the increasing CO_2 concentration. The vegetative growth is often stimulated by increasing temperature as is leaf and canopy photosynthesis. However, reproductive growth leading to seed yield is often depressed by the same increases in temperature. The individual seed growth rates and final mass per seed decline as the temperature exceeds a daily average of about 23 °C (Pan 1996).

Leaf photosynthesis rate increases up to 35 °C and the increase in biomass production was found up to 32 °C (Pan 1996). Pan (1996) demonstrated that canopy photosynthesis and vegetative growth were sustained up to a range of 44/34 °C day/night temperature, respectively. Analysis of long-term data showed that growing season temperatures had a negative impact on soybean yields and caused 17 % reduction for every 1 °C rise in temperature (Lobell and Asner 2003). Putch et al. (2013) observed a linear trend of decline in seed yield components with increase in temperature. They have recorded that the adverse effect of high temperature was greater in a large seeded cultivar of soybean, namely, AGS 190, than that of a small seeded variety (Dieng). Low-temperature stress in colder areas adversely affects the productivity potential of soybean. Ohnishi et al. (2010) reported that pod setting and fertilization are the most vulnerable stages to low-temperature stress. They have demonstrated that low-temperature-induced abnormal pollen grains and reduced number of pollens on the stigma decrease male fertility subsequently affecting pod set and fertilization in soybean. Tacarindua et al. (2013) did not find the significant reduction in biomass production and photosynthetic activity in soybean by increased temperature, but its seed yield was significantly reduced under the temperature higher than 30 °C. They observed that the reduction in the seed growth rate and seed size was due to the reduction in the number of cells per cotyledon.

Jones et al. (1985) in an experiment under phytotron facility observed that evapotranspiration was 20 % more at 31 °C and 30 % greater at 35 °C than that at 28 °C. The transpiration increased about 4 % per 1 °C rise in temperature. Instantaneous and daily canopy evapotranspiration increased drastically as leaf temperature and air vapor pressure deficit increased. The doubled CO_2 decreased canopy evapotranspiration only 9 % at a mean temperature of 23 °C, whereas there was no CO_2 effect on canopy resistance and evapotranspiration, when mean temperature was 35 °C or higher. The study of Allen et al. (2003) on soybean indicated that CO_2-induced water saving by reducing evapotranspiration (due to canopy resistance and stomatal closure) will be considerably offset by the increase in temperature. This change in transpiration will increase the water requirement for soybean crop production. The high temperature did not significantly reduce the photosynthesis, but it did cause lower fertility, higher abortion of pods, and reduced the size of seeds. Water use efficiency in soybean plants was sharply decreased as temperature increases. However, only cool regions would likely benefit from higher temperature. High cost for water, lower yields, and lower WUE could cause problems for the production of soybean in high-temperature areas particularly in arid regions. Studies on soybean indicated hydraulic conductance between the xylem and guard cells of *Glycine soja* which may be controlled by an aquaporin sensitive to aquaporin inhibitor $AgNO_3$ (Sadok and Sinclair 2010). This aquaporin function seems to be temperature sensitive (Lonenko et al. 2010).

However, reproductive growth leading to seed yield is often depressed by the same increase in temperature that improves the vegetative growth. Seed growth and seed weight of soybean reduce as the daily average temperature increases more than 23 °C. Reddy et al. (2002) demonstrated that post-flowering development is a function of thermal units accumulated after flowering. Putch et al. (2013) observed that at the temperature about 30 °C, the yield components were not affected, whereas they were adversely affected at 35 °C. The rise in temperature at flowering and pod set is critical to soybean seed yield. Reduction in seed size under high temperature during seed fill was related to shorter seed growth period. High temperature during reproductive period adversely affected the cell expansion, cotyledon cell numbers, and seed-filling rate resulting into small-sized seeds. The rise in temperature caused an increase in the percentage of aborted pods, lower seed set per pod, increase in number of nonviable pollens, failure of anther dehiscence, and reduced pollen tube penetration in stigma. High temperature degenerates the tapetal layer resulting in abnormal pollen grain formation in the temperature-sensitive soybean cultivars. Salem et al. (2007) concluded from their studies that pollen parameters, i.e., pollen numbers, their growth and development, and pollen tube growth, would be good indicators in determining the tolerance of reproductive phase to high temperature. These pollen parameters may prove more important than growth parameters in screening high-temperature-tolerant soybean genotypes.

Pan's (1996) observations with Bragg variety of soybean indicated that the elevated CO_2 effect on seed yield reaches its peak between 32/22 °C and 36/26 °C daily average day and night temperature, respectively, with a rapid decline at the higher range of temperature. The warming in temperate regions will increase soybean crop production in US soybean growing region (Kucharik and Serbin 2008). Heinemann et al. (2006) observed a significant increase in soybean yield under elevated CO_2 conditions; however, the percentage of flower setting seeds was lower despite their greater photosynthetic rates (Prasad et al. 2002). Prasad et al. (2002, 2005) attributed this to an increase in tissue temperature at elevated CO_2. Currey et al (1995) predicted decreased soybean yields in South Eastern USA associated with 5 °C increase in temperature predicted by various global climate change models. Soybean produces a small amount of viable seeds with decreasing quality as temperature increases above optimum (about 26 °C). Progressive reduction in seed yield occurs with increasing temperature despite the fact that vegetative growth and photosynthesis remains unaffected at much higher temperature (44/34 °C average day and night temperature). The adverse effect of higher temperature on flower initiation, floral development, pollination, fertilization, early embryo development, and seed-filling processes contributes to the temperature-induced reduction in seed yield of soybean.

Vu et al. (1997) in a study on the comparative response of rice and soybean demonstrated that high-temperature treatments reduced the rate of photosynthesis in rice and soybean leaves by 25 and 38 %, respectively. They explained that doubling CO_2 reduced the Rubisco protein content by 22 % in rice but only 8 % for soybean. The elevated growth temperatures (day) above 32 °C also reduced the Rubisco protein content. The Rubisco content declined by 3.8 and 2.2 %, respectively, in rice and soybean leaves for each 1 °C rise in temperature. The soybean leaf Rubisco content was less susceptible to elevated temperature than that of rice. Both the elevated temperature and [CO_2] exert coarse and final control over Rubisco activity, with rice being more susceptible to downregulation of this enzyme than soybean.

4.3.1 Pest and Pest Management

Elevated CO_2 increased the susceptibility of soybean to invasive insects by down-regulating the expression of genes related to hormonal defense. These genes down-regulate important anti-digestive defenses against the beetles. Soybean responds to insect attack by producing defense compounds that inhibit digestive enzymes (proteinases) in the gut of insects, thereby reducing their activities to damage the crop. The production of these anti-digestive compounds is regulated in plants by the hormone jasmonic acid. However, elevated CO_2 levels disrupt this equilibrium in plant-insect interactions and benefit herbivore.

Bonde et al. (2012) showed that increasing temperature contributed to the constraints of soybean rust development. Evidences indicated that high temperature limits soybean rust development due to shortage of moisture on the surface of plants. This would restrict the urediniospore germination and infection. Long-distance urediniospore transport is exposed to sunlight which is detrimental to their viability. This explains how soybean rust fails to develop in the warmer season.

4.3.2 Grain Quality

Allen and Boot (2000) demonstrated that higher growth temperature decreases the yield and quality of soybean seeds. The unique chemical composition of soybean has made it one of the world's most valuable agronomic crops. It is being consumed for several years in Asia and has now been used as food in the USA also (Liu 1997). Its nutrient composition includes 20 and 40 % oil and protein, respectively. Other components of soybean are carbohydrates (30 %), crude fiber (5 %), and ash (5 %) (Thomas et al. 2003). The intrinsic value of soybean seed is in its supply of essential fatty acids and amino acids in oil and protein, respectively. The nutrient composition of soybean is significantly affected by genetic and environmental variables as well as their interactions. Several studies have shown changes in seed composition due to day and night growth temperatures ranging from 15/12 °C to 40/30 °C (Gibson and Mullen 1996; Rebetzke et al. 1996). Vegetative growth processes and seed production of the soybean cultivar "Bragg" were significantly affected drastically by the elevated temperature reducing the weight per seed (Baker et al. 1989). As growth temperature of field-grown soybean increased up to a mean of 28 °C, oil concentration increased and protein concentration decreased from 14 °C to a minimum at 22 °C, at which point it began to increase with temperature up to 28 °C (Piper and Boote 1999). In addition to the oil concentration, the increased temperature also affects the fatty acid composition of the soybean oil. The rise in temperature brought about the increase in oleic acid concentration and reduction in the linoleic acid level in soybean oil (Gibson and Mullen 1996 and Rebetzke et al. 1996). Rebetzke et al. (1996) also demonstrated that the high temperature considerably reduced the ratio of polyunsaturated to monounsaturated fatty acids in soybean oil. Changes in the fatty acid composition, such as the concentration of oleic acid, are associated with nutritional aspects as well as storage and longevity of soybean oil (O'Byrne 1995). Oleic acid is more stable because it is monounsaturated, and with only one double bond, it is less susceptible to oxidation than linolenic acid, which has three double bonds. According to Hungria and Vargas (2000), the higher-temperature-induced reduction in the seed protein concentration could possibly be caused by limited nitrogen fixation due to the vulnerability of rhizobial strains to a higher temperature. Total nonstructural carbohydrates decreased continuously as the temperature increased, indicating that available assimilates are increasingly converted to lipid, protein, or structural components. Lower carbohydrate levels in seeds with increasing temperature would suggest assimilates' availability or transfer into seed was limited as hypothesized by Thorne (1982).

The carbohydrate composition significantly changed with increasing temperature from 18/13 °C to 33/28 °C (day/night) in soybean seeds. Wolf et al. (1982) observed that sucrose concentration increased and stachyose decreased slightly; the glucose, raffinose, and fructose did not change significantly with rising temperature. Thomas et al. (2003) studied the combined effect of temperature and elevated CO_2 on the composition of soybean seeds and found that total soluble sugars and starch decreased as temperature increased from 28/18 °C to 44/34 °C (day/night), while the proportion of soluble sugars to starch decreased, whereas the effect of elevated CO_2 is comparatively negligible. Wang et al. (2012) in a proteomic analysis of the effect of increasing temperature on soybean seed protein revealed 42 stress-responsive spots on the 2 DE gel image and indicated that the high-temperature stress-induced reduction in the storage protein is due to reduction in nitrogen assimilation, reduction of systematicness of cell cycle, alteration of membrane fluid, and energy depletion.

Knowledge of the chemical composition of these seeds should aid in ascertaining how environmental factors affect the nutritional value of soybean seeds. The elevated temperature brings about changes in transcript abundance considerably affecting the composition of seeds. Additional study is necessary to understand the biochemical basis for these phenomena. A search for genetic materials that are more tolerant to high temperatures is required for the sustainable production of soybean. Better screening tools and breeding strategies for developing cultivars suitable to cope with abiotic stresses throughout the growth and development stages are required for soybean to grow better under future climatic conditions. Applied research for stomatal regulation of controlling water loss and management strategies, such as altered planting dates to mitigate high-temperature stress effects, needs to be done using the plant growth models.

The selection of the genetic material of soybean that is more tolerant to the high temperature for seed production is needed. Applied research on management strategies such as altered planting dates needs to be conducted, and the use of plant growth models to identify plant types for higher-temperature conditions is required.

4.4 Maize

Maize is the third most important food crops in terms of energy and protein. This crop is cultivated between 50° N and 40° S. Maize is widely grown in the warmer temperate regions and humid subtropics. It is grown on nearly 100 million hectare area with almost 70 % of its production in developing countries (FAOSTAT 2010). It is one of the most cultivated C4 crop in the world (FAO 2011). According to Leakey et al. (2009), maize is most widely grown as C4 grain crop in both food secure and food insecure regions. Thus, the adaptations in maize to higher temperatures would likely to be a priority agenda for policy makers, scientists, and researchers for food security of these regions. The demand of the maize will double by 2050 in the developing world, and it is predicted to become the crop with the greatest production globally and in developing world by 2025 (Rosegrant et al. 2008). According to IPCC report (2007), the maize growing areas of sub-Saharan Africa will encounter increased growing season temperatures and intermittent frequency of drought. Cairns et al. (2013) observed that the maize yield in Eastern and Southern Africa has reduced to 1.4 t/ha^{-1} about one sixth of the average yield in the USA (FAOSTAT 2010). These yield levels are not sufficient enough to ensure food security of African region. The need to increase the yield of maize for food security has become a priority program to meet the challenge of explosive population growth and the drastic threat of climate change.

Maize is grown throughout the year as summer-rainy (Kharif) and winter season's crop (Rabi maize) in many Asian countries. In India, the main maize growing season is Kharif, which accounts for about 85 % of the total maize area in the country. However, in spite of maximum share in the area, the relative contribution of Kharif maize is comparatively much lower than Rabi maize. Kharif maize crop grown in tropical/subtropical environments occasionally faces extreme climatic conditions with various abiotic and biotic constraints, which limit the production and productivity of Kharif maize. Maize is grown at temperatures closer to its upper limit. Progressive temperature increases associated with climate change increase the risk of temperature stress for this crop. However, in much of the European regions and North America, maize is grown in higher latitudes, and the risk of increasing temperature is much reduced. At higher elevations, the rise in temperature may increase photosynthetic efficiency during cooler periods, improving the economic viability of this crop in temperate climates (Wittwer 1995). Relatively less effort is made to study the heat tolerance in maize in comparison to drought stress.

Increasing temperature in general results in large yield losses in maize production in sub-Saharan Africa (Rowhani et al. 2011). Cairns et al. (2013) in an exclusive study on maize indicated that current maize germplasm developed for drought tolerance may not perform well under drought stress at elevated temperatures. The results of this study suggest that the genetic control of drought, heat and combined drought, and heat tolerance are largely independent of each other. Therefore, they suggested that maize breeding for tolerance to drought stress under higher temperatures must include screening under the combined effect of drought and heat rather than the individual stresses.

4.4.1 Phenology

The suitable temperature for the germination of maize is between 18 and 21 °C; however, it fails to germinate below 10 °C. Flowering in maize plant is temperature and day length dependent. The photosynthesis in C4 crop like maize is more tolerant to high temperature due to the absence of photorespiration, which increases rapidly with temperature. However, the maize photosynthesis declines with the temperature above 35 °C. Moreover, the photosynthetic efficiency of maize plants may increase with increasing temperature during the winter season.

High soil temperature above 35 °C decreases the leaf elongation rate, leaf area, shoot biomass, and the rate of photosynthesis. Elongation of the first internode and the overall shoot growth of maize are the most vulnerable processes affected by high temperature. Although C4 plants have a higher optimum temperature for photosynthesis compared to C3 plants, C4 pathway alone did not necessarily confer tolerance to high temperature (Ghosh et al. 1989). Differences in photosynthetic response were more closely associated with light reactions, particularly the sensitivity of photosystem II activity at higher-temperature stress. Given the different traits and mechanisms are likely to provide adaptation for different types of stress (i.e., varying in duration, intensity, and timing), heat stress environments need to be defined to identify and generate the adaptation strategies.

High-temperature stress is another major abiotic stress that reduces the maize grain yield more than 15 %. Invariably high temperature during spring season causes an array of morphophysiological, biochemical, and anatomical changes, which affect plant growth and development more at reproductive stage through pollen abortion, silk desiccation, and reduced grain set leading to a drastic reduction in grain yield. The evapotranspiration in maize increases during heat stress. The higher carboxylation efficiency of PEP carboxylase in comparison with Rubisco in maize facilitates the maintenance of a much higher diffusive gradient for intercellular CO_2 between the atmosphere and the site of carboxylation. This enables the maize plant to maintain optimal internal CO_2 concentration at minimum stomatal conductance, so that the rate of increase in evapotranspiration with the rise in temperature is much reduced. However, the increasing temperature also shortens the crop growth duration and grain yield decreasing its water use efficiency. Maize growth in India is affected adversely when the temperature increases beyond 32 °C. Net photosynthesis is inhibited at temperature above 38 °C due to thermal inactivation of enzymes. The activation state of Rubisco decreases at the temperature exceeding 32.5 °C with nearly complete inactivation at 45 °C. High-temperature stress and low humidity can desiccate exposed silk and pollen grains due to their thin outer membrane when these are released from anthers. The degree of damage depends on the intensity and duration of high-temperature spell (Singh et al. 2011).

Studies in CIMMYT using biotechnology tools demonstrated that most of the morphological parameters were complex and regulated by several genes. The study on physiological parameters such as relative water content (RWC), osmotic adjustment, root conductivity, and chlorophyll content showed that on chromosome 6, a QTL for RWC corresponds exactly to a QTL for ASI and grain yield under stress. The identification of dehydrin gene (dhn1) was also reported in the same chromosome region (Jiang and Zeng 1995; Campbell and Close 1997).

High-temperature stress also adversely affects seedling and vegetative stages. Seedling growth at soil temperature above 26 °C reduces root and shoot biomass by 10 % for each degree increase until 35 °C, when growth is severely retarded (Walker 1969). Riley (1981) attributed the high temperature induced reduced seedling growth to poor reserve mobilization and reduced protein synthesis. Correlation studies in maize (Liu et al. 2011) indicated that leaf temperature response played an integral role in maize biomass accumulation. The leaf temperature difference (LTD), drought tolerance index (DTI) of shoot fresh weight (SFW), and shoot dry weight (SDW) were the traits evaluated for QTL (quantitative loci) analysis in maize seedlings. A total of 9 QTL were detected by composite interval mapping (CIM) for the three traits (LTD, RSFW, and RSDW). Two co-locations responsible for both RSFW and RSDW were detected on chromosomes 1 and 2, respectively, which showed common signs with their trait correlations. Another co-location was detected on chromosome 9 between LTD and shoot biomass, which provided genetic evidence that leaf temperature affects biomass accumulation.

4.4.2 Effect of Cold Stress

Maize is thermophilic plant species, which requires high optimum temperature for germination, growth and development, flowering, and grain growth. Rabi maize faces low-temperature stress during the month of December and January, when the temperature goes as low as 4 °C. There are two types of injuries caused by low temperature:

1. Chilling injury in the temperature range of 0–15 °C resulting into physiological disruptions in germination, flower and seed development, and grain yield and storage life
2. Freezing injury in the external temperature below freezing point causing intracellular (damaging the protoplasmic structures) as well as extracellular (dehydration of the cell) freezing damages.

However, the major limiting factor for winter maize cultivation is the chilling injury. Chilling at germination damages the cells at imbibition. The low temperature at growing period drastically damages the process of chlorophyll formation in leaves resulting in lighter-colored leaves, whereas chilling at later stages results into the brown-colored leaves. The effect of chilling injury may go in a long way resulting into a drastic reduction in grain yield and poor-quality grains.

The minimum temperature for shoot development in maize is 32 °C. Low-temperature stress resulted in reduced plant height. A reduction of 30–40 % was observed in inbreds and 15–25 % in hybrids. The optimum temperature for leaf expansion in maize plants ranged from 7 to 30 °C (Tollenar et al. 1979). However, the temperature below 10 °C decreases leaf expansion and kills shoot apical meristem (Miedema 1982).

Less attention has been paid for flowering and grain-filling cold tolerance, due to the fact that these stages generally come during February/March in North Indian cultivation of winter maize. However, with changing pattern of cropping systems and global climate changes, there are chances of the exposure of maize plants to the low temperature at flowering/grain-filling stages, which are highly susceptible to cold stress. Controlled environmental studies showed a reduction in a number of tassel branches and germination of pollen grains under low temperature (Bechoux et al. 2000). Fast early growth and early flowering with long grain-filling duration and extended stay-green character provided the basis for hybrid development with high yield potential under low-temperature stress (Frei 2000).

Maize can survive when exposed to adverse temperatures as low as 0 °C and as high as 44 °C. Lukatin and Naraikina (2011) observed that unfavorable temperatures, i.e., below 3 °C (low) and above 44 °C (high), cause inhibition of growth of axial organs, intensification of lipid per oxidation, and reduction of antioxidant activity by which the intensity of high-temperature stress could be identified. The degree of antioxidant activity determines the negative effect of temperature stress on maize seedlings. Growth starts slowing down when the temperature dips below 4.9 °C. The low-temperature stress during vegetative phase results in the reduction in plant height, yellowing of leaves, chlorosis, and leaf tip burning due to the death of leaf tissue. Cold stress at reproductive phase severely affects flowering resulting in reduced tassel size/branches, delayed anthesis, pollen grain death, reduced silk size, and seed setting. In northern hill region of Himalaya, temperate climate exists. Rising temperature will increase the rate of germination, seedling growth, and viability of maize. The minimum and optimum temperatures for the germination of maize are lower than sorghum. The cardinal temperatures for maize germination are about 3–10 °C below those of sorghum. The rate of development in maize may be assessed in growing degree days (GDD). According to Roth and Yocum (1997), the maturation of short-, medium-, and long-season maize hybrids of North America requires 2100, 2400, and 2800 °C growing degree days, respectively.

4.4.3 Grain Quality

Lu et al. (2014) in an experiment on the effect of heat stress at grain-filling phase demonstrated that heat stress decreased the number and weight of grains. Starch accumulation was decreased when the temperature was increased between 1 and 10 days after pollination. However, heat stress at other time did not affect the starch deposition in maize grains. The iodine-binding capacities of starch were increased due to elevated temperature. The peak and breakdown viscosities and gelatinization enthalpy were decreased. However, the gelatinization and pasting temperature increased. These changes in the physicochemical properties of starch and protein as well as in the long chains in amylopectin affect the pasting and thermal properties of maize grains. High temperature reduces kernel size by decreasing both the accumulation of dry matter and duration of grain fill. It lowered the shelling percentage, reduced test weight and grain oil, but increased the grain protein content. The maize protein quality is poor due to deficiencies of essential amino acids lysine and tryptophan. Maize grain protein consists of 3 % albumin, 3 % globulin, 60 % prolamin (zein), and 34 % glutelin. Waxy maize contains 100 % amylopectin starch, the high amylose maize has 40–70 % amylose content, and amylomaize is having amylose as high as 85 %. The oil deposition in maize starts late during seed development and decreases with high-temperature exposure resulting in reduced oil content. The fatty acid composition also gets influenced by production of saturated fatty acids under warmer conditions, and the warming increases the percentage of monounsaturated fatty acid (oleic acid) and decreases polyunsaturated fatty acid (linolenic acid) (Seiler 1983).

4.4.4 Model

Most widely used models for predicting the impact of temperature in maize crop are CERES-Maize and EPIC and a recent maize simulation model CSM-IXIM, which describes the individual leaf area growth, leaf level of carbon assimilation, partitioning to canopy level, and growth of reproductive organs. This studies a simple feasible option for improving GDD-based crop models to integrate nonlinear temperature relationships without sacrificing their current structure (Kim et al. 2012). However, according to Cicchino et al. (2010), prediction of phenology for maize crop needs to be based on thermal time (TT), computation of accurate Basel (Tb), and optimum temperatures (To). By the end of this century, growing season temperatures will exceed the most extreme seasonal temperatures recorded in the past century (Battisti and Naylor 2009). According to Dale (1983), the maize yields to be negatively correlated with accumulated degrees of daily maximum temperatures above 32 °C during grain-filling period in the US Corn Belt. Lobell and Burke (2010) suggested that an increase in temperature of 2 °C would result in a greater reduction in maize yields within sub-Saharan Africa than a decrease in precipitation by 20 %. Lobell et al. (2011) showed that for every degree day above 30 °C, grain yield was reduced by 1–1.7 % under optimal rain fed and drought, respectively. The temperature threshold for damage by heat stress is significantly lower in reproductive organs than in other parts (Stone 2001). According to Maiorano and Mancini (2010), temperature stress effect starts when the developing grain's moisture level poured down below 30 %; subsequently water activity increases under high temperature bringing a deleterious effect on maize grain development. High temperature during reproductive phase is associated with a decrease in yield due to a decrease in the number of grains and kernel weight. The number of ovules that are fertilized and developed into grains decreases under high temperature (Schoper et al. 1987a, b). Dupuis and Dumas (1990) demonstrated that female tissues have greater tolerance; however, the pollen produced under high temperature has reduced viability and in vitro germination. They have also observed that the high temperatures are responsible for reduced pollen water potential, the quantity of the pollen shed, and pollen tube germination. High temperature at early kernel development adversely affects the final kernel mass due to the reduction in the number and size of endosperm cells reducing the sink potential of maize cultivars. The high temperature during grain-filling period reduces the duration of grain filling (Muchow 1990). Wilhelm et al. (1999) attributed the heat-induced reduction in seed mass to the greater reduction in the rate of grain filling. They also suggested that the heat stress adversely affects the conversion of sugars to the storage product starch. Enzymatic studies (in vitro) on the effects of high temperature on the carbohydrates in maize kernels suggest ADP glucose pyrophosphorylase and sucrose synthase were most vulnerable to high-temperature stress affecting the starch accumulation adversely. The kernel dry mass is generally maximum at 22 °C and declined linearly with the rise in temperature to 36 °C, where seed weight was reduced by 45 %. This reduction was due to the greater reduction in the activities of ADP glucose pyrophosphorylase and starch synthase in the developing kernel at a higher temperature (Singletary et al. 1994). Thomson (1966) showed that an increase in temperature from 22 to 28 °C during grain-filling period in US Corn Belt resulted in a 10 % yield loss. A recent study on 20,000 maize trials in Southern Africa demonstrated the linear reduction in maize production with every accumulated degree day above 30 °C (Lobell et al. 2011). Heat stress in maize is associated with shortened life cycle, reduced light interception, increased respiration, reduced photosynthesis, and pollen sterility. However, in a recent study Cicchino et al. (2010) showed the vulnerability of the period between silk pollination and ovary fertilization to heat stress also contributes for the determination of the yield in maize plants. It was demonstrated by Lafitte et al. (1997) that tropical maize is less vulnerable to temperature stress than that of temperate maize. Tao and Zhang (2011) in a study in China on the impact of global mean temperature on maize productivity indicated that the decrease in yield due to 1 °C, 2 °C, and 3 °C rise in global mean temperature (GMT) ranged from 1.4 to 10.9 %, 9.8 to 21.7 %, and 4.3 to 32.1 %, respectively. However, the yield losses were larger for rain-fed maize.

4.4.5 Plant Diseases

Changes in temperature, humidity, and rainfall pattern have the potential to increase infection by many maize pathogens. Increased temperature reduces the generation time resulting in a higher number of generations per season. It accelerates and increases the inoculum loads affecting pathogen evolution rates and its capacity to adapt faster to the environment than the host. The wet weather along with the cold temperature favors the activity and development of soil pathogens which initiates the killing of germinating seeds. The mycotoxins occur in maize system because pathogens contaminate agricultural products and threaten food security. It is a serious problem with long-term consequences for human and animal health. Fungi such as *A. flavus* and *A. parasiticus* produce aflatoxin, *F. verticillioides* produces fumonisin, and *F. graminearum* produces deoxynivalenol (DON). Consumption of high doses of mycotoxins causes acute illness and can prove fatal. Increasing temperature in maize growing regions likely changes the distribution and predominance of *Fusarium verticillioides* where it will replace *Fusarium graminearum* causing increased incidence of fumonisin contamination (Miller 2001; Torres et al. 2007).

4.4.6 Pest Management

InfoCrop maize, coupled with holometabolous population dynamic model, was used to simulate population dynamics of maize stem borer, *Chilo partellus*, as well as crop-pest interactions. Maize stem borer acts as a stand reducer and causes loss of leaf area, leaf weight, stem weight, and panicle weight to the crop. Due to larva being the damaging stage of the pest, the larval population was linked to the processes of leaf area, leaf weight, stem weight, and ear weight in the crop model. Depending upon larval population and feeding rate of larva, these growth processes were affected. Simulation of pest dynamics showed a decline in the pest severity, thereby reducing the pest-induced yield losses under global warming. However, despite the reduction in pest stress, maize crop productivity as such may be adversely affected by the rise in temperature.

Temperature is the most important factor influencing insect development. Phadke and Ghai (1994) reported that a global rise in temperature accelerates development of insects from egg to adult, leading to increase in population, more crop damage, and yield loss. The accumulated degree days provides the physiological time in which insect development progress fairly (Kingsolver 1989). GCMs indicate that future temperature changes will be greatest in winter. Higher winter temperature will increase the survival of many overwintering species such as corn earworm (*Heliothis zea*) in the Northern Great Plains of North America leading to greater damage of maize and soybean crops. As the temperature changes, the impact of increased overwintering may be greater at high altitudes. Temperature is less limiting in tropics and subtropics. Higher temperatures result in the faster development rates and a more rapid increase of pest populations as the time required for reproductive maturity is reduced (EPA 1989). The incubation period of *Chilo partellus* egg was 9 days at 56–60 °F, and with the increase in temperature (80–90 °F), it came down to 4 days. Similarly pupation did not occur below 55 °F, above which the percentage of pupation increased progressively with the rise in temperature. High proportion of pupation occurred at 71–75 °F but showed progressive decrease with further increase in temperature up to 96 °F, where pupation was completely ceased (Chaudhary et al. 2005). Higher temperature generally results in faster rates of development and shorter development time. Thus, increase in the number of generations of *Heliothis zea* on maize crop is expected. Wiseman and Isenhour (1989) reported that the growth of this pest was slow at 20 °C and relatively faster at 30 °C, while 25 °C is optimum for their growth. Thus, at a higher temperature, number of generations and higher population to start additional generations increase infestation (Furher and Gregory 2014). According to Mattson and Hoack (1987), the increase in temperature affects pest-parasite interaction in such a manner that minor pests become major affecter as the case of *Pyrilla perpusilla*, which became the severe attacker in maize. The increase in temperature may allow the crop to migrate to colder regions resulting into the expansion of both crop and pests into those areas. The outbreak of pests is weather related and occurs with the stressed plants where the defense system and resistance are lowered leading to more infestation and pest outbreaks.

4.4.7 Weeds

The problem of weeds is relatively more in maize crop grown during summer-rainy season (Kharif). The hot and humid condition during Kharif is highly congenial for weed growth and development. Sahoo and Saraswat (1988) reported 45 % reduction in maize yield due to unchecked weed growth. However, in the case of weeds, genetic improvement option has a least opportunity; reducing losses due to weeds is largely based on chemical, mechanical, and manual weed control measures. *Cynodon dactylon*, *Echinocola*, *Amaranthus*, and *Cyperus rotundus* are the major weeds affecting maize production. Altering planting date, competitive crops, crop rotation, intercropping, selective stimulation of crops, and chemical or manual weeding help in minimizing the crop yield losses.

4.4.8 Mitigation Techniques for Low-Temperature Stress

The early maturing hybrids were affected due to cold for height at vegetative stage, whereas medium maturity single-cross hybrids tolerated prolonged cold stress. Thus, during cold stress conditions, medium maturing full-season cultivars and single-cross hybrids are recommended for general cultivation due to their cold stress escaping nature.

1. Planting of seeds on the southern side of the ridge to get maximum sunlight during the daytime. Planting the tall barrier crops on the northern side will help to protect growing maize plants from cold winds.
2. Potassium application helps in protecting the roots from the cold stress.
3. Frequent irrigation to maintain a conducive temperature in the rhizosphere.
4. Growing of medium maturity single-cross hybrids, which are more cold tolerant than open pollinated cultivars.
5. Selection of genotypes with dark green, purple color and erect leaves and high density lateral roots.
6. ABA-induced somaclonal variants showed high accumulation of proline, resulted in increased cold tolerance.

4.4.9 Mitigation Technologies for High-Temperature Stress

1. Selection of early maturing single-cross hybrids.
2. Planting of seeds in furrows.
3. Sprinkler irrigation system reduces the high-temperature effect in plants.
4. Change the sowing time to avoid high-temperature stress at flowering.
5. Selection of cultivars/hybrids which have shorter anthesis-silking interval.
6. Organic wastes and mulching of soil help in conserving the soil moisture to mitigate the heat stress.
7. Staggered sowing in larger growing area prolongs the pollen availability throughout the flowering period.

4.4.10 Future Thrust

Global warming will make current maize-producing regions highly vulnerable. However, as its cultivation shifts toward higher altitudes, the crop faces low-temperature stress leading to poor germination, photoinhibition of photosynthesis, and reduced water and nutrient uptake by seedlings. Thus, the photoinsensitive maize varieties will be required to adapt to this change in climate as the temperature rises. Rapid advances in methods for adapting the maize crop to lower temperature have been achieved by plant breeders. Genetic transformation to enhance protection against active oxygen radicals may further aid in this effort. The increased evaporative demand and increased incidences of supraoptimal or undesirable temperatures may make large areas of subtropical regions unsuitable for maize cultivation necessitating for identifying temperature-tolerant maize races and redesigning the current cultivars to changing climatic conditions using new tools of biotechnology to ensure future maize yield stability.

Maize plants in farmer's fields are routinely subjected to a combination of stresses. The occurrence of drought and heat stress together is likely to increase particularly in Southern Africa. The study by Cairns et al. (2013) suggested that the genetic control of drought and heat and a combination of both these stresses are largely independent of each other. This independent control indicates that quantitative trait loci (QTL) with large effects are involved in combined drought and heat tolerance. Farmers have adapted themselves to the impact of climate variability; however, climate change represents a greater challenge. Adaptation to climate change requires effective policies and management strategies at the country, regional and international levels to ensure the technologies reach the intended beneficiaries and make the desired impact. Newly developed varieties/hybrids with increased resistance to high-temperature stress will play a significant role in adaptation to climate change (Easterling et al. 2007). While the challenge is immense, the advancement of molecular tools and knowledge of physiological mechanisms responsible for yield loss will provide solid adaptation strategy to meet the challenge of high-temperature stress.

4.5 Cotton

Cotton (*Gossypium hirsutum* L.) is a fiber crop, which stands fifth in the line of world's most economically important crop. It is a perennial plant with complex growth habit. It is highly sensitive to adverse environmental conditions. Its area of cultivation was over 30.0 M ha in the world during the period 2011–2012. Most of the world's cotton cultivation area falls in arid and semiarid climatic regions. The cotton growing area in India during this period ranges more than 12.0 M ha (Hebber et al. 2013). The primary factor affecting growth and development in the cotton plant is the temperature. Cotton is grown in a relatively narrow temperature range. The minimum temperature for growth and development of cotton is 12–15 °C, the optimum temperature is 26–28 °C, and maximum temperature depends on the duration of exposure (Reddy et al. 1997). Even a short period of above optimum canopy temperature may cause injury to young fruits. Cotton canopy temperature is generally cooler than air temperature (Kimball et al. 1992a, b).

The ideal temperature for the growth and biomass accumulation ranges between 20 and 30 °C. The optimum range of temperature for the physiological and biochemical processes in cotton plants is estimated to be within 23.5 and 32 °C. General circulation models predict the 3.90, 3.20, and 1.85 °C rise in the mean temperature of main cotton growing regions of India, respectively. Hebber et al. (2013) reported that according to InfoCrop cotton model, the cotton seed yield will be reduced by 477 Kg ha^{-1} in most vulnerable regions followed by a decline of 268 Kg ha^{-1} in the less susceptible zone. The yield decline due to increasing temperature and reduced precipitation was greater in northern zone compared to the southern region.

4.5.1 Growth and Development

The variability in temperature controls the growth and developmental processes such as organ initiation (leaf, flower node) and the time interval between flowering and fruit maturation. According to Reddy et al. (1991a, b), photosynthetic rates are less sensitive to temperature than are growth and development rates in cotton. Accordingly, growth is not considered a limiting factor controlling photosynthesis in cotton. Hodges et al. (1993) mentioned that the computer-based crop model and expert system GOSSYM-COMAX is the first crop simulator that can determine the cotton crop responses to weather, soil, and management conditions. It describes that the temperature optimum for stem and leaf growth as well as seedling and fruiting in cotton is about 86 °F. The effect of rising temperature varies markedly between cultivars for the developmental processes such as the rate of node formation, days to first square, first bloom, and open boll as well as flowering. They also described that the fruit retention decreased rapidly as the time of exposure to above 104 °F increased. However, when natural temperature conditions exceed the temperature information from which the model was developed, it does not accurately predict growth under either low or high temperature. It was shown by Baker and Allen (1993a, b) that the apparent photosynthesis was relatively insensitive to air temperatures between 20 and 40 °C. However, conditions that induce different growth rates resulted in modified photosynthetic rates. Photosynthesis was much greater in rapidly growing plants. Developing seedlings translocate about 50 % of their photosynthates into roots, whereas mature cotton plants had only 5–10 % of their dry mass in roots. After the growth of stem and leaves up to 3 weeks, most of the photosynthate was utilized for the development of bolls and little translocated to roots or leaves and stem. This translocation of photosynthates was strongly affected by temperature.

According to Rawson (1992), cotton is a high-temperature C3 crop, grown in summer with the capacity to branch; its shoot dry weight increased significantly by CO_2 enrichment only at higher temperature. Yoon et al. (2009) demonstrated that by increasing CO_2 level of the area to 600 umol/l^{-1}, the total biomass of cotton plants increased at lower (25/15 °C) as well as higher (35/25 °C) day/night temperatures; however, a further increase of CO_2 up to 800 umol/l^{-1} increased total dry matter production only at higher-temperature condition. They have also observed that at higher temperature elevated CO_2 will have more positive effect toward the end of growing season. At high-temperature cotton acted as it was source limited. It appeared to have spare sink capacity which CO_2 could partially satisfy, even under the higher irradiance. The cotton during early boll filling depends on short-term stored assimilates. The increased temperature increases the rate of development and concurrently reduces the number of days during which the radiation can be intercepted and the plant becomes more source limited. CO_2 enrichment will then be more likely to enhance yield by filling the role of radiation.

High night temperature is one of the key factors causing the marked reduction in cotton yield. Loka and Oosterhuis (2010) attributed this to the significant increase in the rate of respiration and reduction in carbohydrate accumulation. They observed the increase of 49 % and 56 % in respiration rate at high night temperatures of 27 and 30 °C, respectively. The ATP levels were proportionately decreased (38 % and 37 %, respectively) following the sequential rise in temperature. Similarly, high night temperature brought about 64–70 % reductions in sucrose and 28–39 % reduction in hexoses in cotton plants. Zeng et al. (2012) indicated that the respiratory metabolism in the reciprocal hybrid of cotton was markedly greater than that of their parents under high-temperature conditions. They have observed that stomatal conductance in cotton leaves was more sensitive to the elevated temperature than the rate of photosynthesis. They concluded that the impact of higher leaf temperature on opening and closing of stomata is more important in determining the photosynthesis than the intercellular concentration of CO_2. The amelioration of high-temperature effects can be possible by developing heterosis in the net photosynthesis in the reciprocal hybrid of cotton which may be useful for the adverse environmental conditions such as high light intensity (>1000 umol m^{-2} S^{-1}), elevated temperature (> 40 °C), and relative humidity (<40 %). High night temperature is also considered to be responsible for increased fruit shedding and decreased boll setting.

Cotton yields are a function of growth, flower production, and flower and boll retention and a summation of individual boll growth rates during fruiting period. The developmental processes in cotton plants including floral initiation, flowering, boll opening, and leaf initiation are relatively sensitive to higher temperature. These developmental processes are highly temperature dependent, because rise in atmospheric temperature or little rise in canopy temperature is generally reflected in overall development of cotton plants. The temperature above 30 °C caused young bolls to abscise. Doubling CO_2 did not ameliorate the adverse effect of high temperatures on cotton fruit retention. Reddy et al. (1991a, b)) found that most of the squares and flowers of the cotton were aborted above 30 °C. Reddy et al. (1992) observed that exposure of cotton plants to 40 °C for 2 h d^{-1} retains 100 % of their bolls by the time 10–11 bolls were on the plant, whereas similar plants exposed to 40 °C for 12 h d^{-1} retained 0 % bolls by the time 3 bolls had been formed. Cotton plants tolerated 40 °C temperature up to 6 h d^{-1} with little influence on boll set; however, exposure of similar plants for 12 h d^{-1} caused complete failure of the crop to produce new squares and fruiting structures.

Kimball et al. (1992a, b) in a study on the measurements of foliage temperature using infrared thermometer of cotton plants observed that ball temperature was about 1.9 °C more than leaf temperature at comparable positions and illumination levels in the canopy. It was observed that CO_2 enrichment increased foliage temperature by an average of 0.80 °C above the ambient CO_2 plot. The rise in temperature from 26 to 36 °C linearly increased transpiration rates per unit leaf area (Reddy et al. 1998). Allen et al. (2003) demonstrated that high temperature did not significantly reduce photosynthesis, but it did cause lower fertility, higher abortion of pods, and smaller shriveled seeds. Water use efficiency on the basis of seed yield would decrease sharply as temperature increases. Elevated temperature would also cause a greater water requirement for plant growth; thus, only temperate area with cooled regions would likely benefit from such rise in temperature. Their observations clearly indicate that under high-temperature conditions, the small savings in evapotranspiration associated with stomatal closure will be considerably offset by greater rise in temperature, which could increase the requirement of water for crop production. They demonstrated that developmental activities occurred more rapidly as temperature increased. The number of days to the appearance of flower bud, first flower, and mature open balls decreased as the temperature increased during the development of cotton plants. Reddy et al. (1997) reported that the time required to produce flower buds, flowers, and mature fruits was reduced by an average of 1.6, 3.1, and 6.9 days per degree of increase in temperature, respectively. Much of the reduction in developmental time occurs during the boll growth period, and this resulted in smaller bolls, lower yields, and poor-quality lints (Hodges et al. 1993). The fruit production efficiency increases as the temperature increased to 29 °C; however, if temperature increases more than this critical temperature, the fruit retention declines significantly. More than critical maximum temperature of 32 °C bolls may abscise prematurely earlier than the normal duration; however, flowers, which develop at cooler temperature, do not abscise even when the temperature rises more than 32 °C. There is a short period during flowering when cotton is highly vulnerable to the elevated temperature, and if developing bolls escape this period, the adverse effect of high temperature could be mitigated (Reddy et al. 1998). Tian et al. (2014) indicated that growing degree days play a key role in determining the quality of cotton fiber. They reported that the cellulose deposition rate was significantly correlated with growing degree days and daily minimum temperature; however, the onset of cellular deposition was significantly affected by growing degree days and daily maximum temperature. The cellulose deposition rate can be increased by warmer daily minimum temperature. Warmer daily maximum temperature advanced the onset of rapid cellulose deposition.

Lu and Zeiger (1994) indicated that one of the options for mitigating the high-temperature stress on cotton plants is the transpirational cooling. Prediction of high temperature may instigate farmers to modify their existing cultivation practices by planting earlier to escape the adverse effects of high temperatures by completing flowering prior to the onset of high temperatures. There are also great chances of shifting of geographical distribution of cotton cultivars toward more productive marginally cool areas. Lu and Zeiger (1994) observed that some varieties of susceptible Pima cotton species escape the effect of high-temperature stress by transpirational cooling. The young bolls of most of the cotton varieties are vulnerable to high temperature. The increase in crop tolerance to elevated temperature and short-term heat shock would be useful to sustain cotton crop productivity in tropical and subtropical regions. The heat-tolerant cotton cultivars may be screened selected and constructed by biotechnology tools to meet the challenge of high-temperature stress. Conaty et al. (2012) suggested that mitigation of high-temperature stress effect on crop plants is possible by rescheduling the irrigation according to the need of plant's growth and development to maximize the performance of cotton plants. Irrigation during the physiological phenophases, when cotton plants are highly vulnerable to an elevated temperature (<28 °C), is desired to counter high-temperature stress effect.

4.6 *Brassica*

Brassica is one of the important oilseed crops contributing about 30 % of total oil production in India. The rapeseed-mustard constitutes an important group of oilseed *Brassica* crop, and India is the second-largest cultivator of mustard after China. However, the country has now recorded a decrease of 6.07 million tonnes (MT) of total oilseed production during 2014–2015 as compared to 2013–2014; a large part of this reduction has been contributed by the rise in temperature and reduction of 20,000 ha of the land under oilseed growing area. Another contributor to this reduction in its productivity is that the 72 % of *Brassica* growing area is confined to rain-fed farming. The oilseed *Brassica* comprise of four species *B. campestris*, *B. juncea*, *B. napus*, and *B. carinata*. Together they occupy more than 14 million hectares with an annual production of 24.61 million metric tonnes (Anonymous 1991). Basically, oilseed *Brassica* is a crop of temperate region. However, due to shortages in edible oil production, mustard cultivation is being extended to nontraditional areas.

Mustard is much sensitive to climate variables such as temperature changes. According to Boomiraj et al. (2010), the yield reduction is expected to be higher in Eastern India (67 and 57 %), followed by Central India (48 and 14 %) and northern India (40.3 and 21.4 %). Their study predicts an overall negative effect on mustard farming from 2020 through 2050 till 2080. The maximum increase in temperature is expected in Eastern India. Indian mustard (*B. juncea*) is an important edible oil-yielding crop accounting for about 80 % of the cultivated area in northwestern part of India (Singh et al. 2014). Among all the states of India, Rajasthan is an important producer of Indian mustard. Indian mustard is very sensitive to heat stress at an early seedling stage. The annual requirement of this oilseed is much higher than that of its production. The low production in the *Brassica* crop is due to:

1. Larger *Brassica* growing area is rain fed.
2. Low resource capacity of farmers for its cultivation.
3. Low technology available for its production.
4. Vulnerability of *Brassica* crop to changing climate.

The seed yield of *Brassica* plant is determined by a number of pods, seeds per pod, and seed weight (Mc Gregor 1981). The number of pods is determined by initiation of flower buds and is dependent on the plant biomass and combined stress levels at that time. Seeds per pod depend on photosynthate supply during fertilization. The seed weight depends on the continuous supply of photosynthates after fertilization and during maturity. However, the reproductive phase has been identified as one of the most vulnerable stages for temperature stress. Singh et al. (2008) considered pollen-based parameters in determining the cold and heat tolerance for winter-grown canola (*Brassica napus*) plants. Their study demonstrated that average cardinal temperatures, i.e., T min., T opt, and T max of pollen germination and pollen tube growth, were 6.4, 24.3, and 33.7 °C, respectively. The study showed that T min and T opt of pollen germination and pollen tube growth help to determine cold tolerance, whereas T max for pollen germination and pollen tube growth helps in determining the heat tolerance in canola cultivars. Canola cultivar KS 3077 as cold tolerant and Kadore as heat tolerant were selected on the basis of their pollen-based parameters.

Uprety et al. (1996) described that *Brassica* crop in India stays in the field from September/October to February/March. Good cooling weather and adequate moisture in the soil during the vegetative, flowering, and seed-filling stages are most critical to realize optimum production. Development of rapeseed-mustard crop may be characterized in terms of the timing of major phenological events such as seedling emergence, floral initiation, stem elongation, appearance of first flower, cessation of flowering, and maturity. Each of these developmental phases affects the production of *Brassica* yield. Sastry and Kumar (1981) reported that cultivar Pusa Bold (*B. juncea*) took 52, 95 and 126 days, respectively, to attain 50 % flowering, cessation of flowering, and maturity in Eastern Uttar Pradesh, while under the Delhi condition, the same cultivar took 52, 77 and 130 days to attain the respective stages. The date of sowing significantly affects the productivity of *Brassica* crop. Leach et al. (1994)

recorded significantly greater yield of *B. napus* when sown in early September compared to those sown in mid-August. Similarly in *B. juncea* 15–60 days delay of sowing reduced the seed yield by 12–66 % (Saini 1984). Tomar and Mishra (1991) observed that early sowing (30th of October) resulted in significantly higher yield of *B. juncea* than that sown in later dates. The reproductive stage was comparatively longer, in early sowing, which resulted in higher seed yield. Suboptimal temperature can also delay seeding reducing the potential yield in that growing season. Prasad (1989) observed that the duration of pod filling ranged from 22 to 33 days depending on the date of sowing and season. The number of days taken to the maturity of the crop varied from 127 days for late sown crop to 137 days for the early sown crop during relatively warm season, when the daily mean temperature ranged from 13.4 to 28.4 °C, whereas the same variety took 138–152 days in comparatively cooler season when the mean daily temperature ranged from 10.1 to 26.7 °C. The maximum leaf area index (LAI) occurred between 70 and 85 days after sowing, and the peak value of LAI ranged from 6.2 for early October sown crop to 1.3 in late November sown crop (Ravindra 1985). The yield reduction due to sowing was found to be less in *B. campestris* (Toria-T9) (1–14.5 %) than in *B. napus* (B.O.–54) (6.7–33.7 %) and *B. juncea* (Pusa Bold) (18.4–40.8 %). The higher temperature was found to have led to a decrease in the duration of reproductive phase resulting in a considerable reduction in total dry weight and seed yield. Higher temperature decreases the rate of photosynthesis at the flower initiation stage, and this has resulted into the poor production of pods and seeds with lesser seed weight. Although maximum temperature stress is experienced during pod formation stage, the susceptibility of *B. campestris*, *B. napus*, and *B. carinata* was greater at critical flowering period (Uprety and Tomar 1993; Uprety et al. 1990). It is thus clear that these phenological events significantly affect the productivity of *Brassica* crop and any change in temperature during these stages markedly affects the seed yield of *Brassica* species. These Indian studies envisage that with the climate variability we have in the *Brassica* growing area of Indian subcontinent, i.e., variations in temperature and other climate variables, the production of *Brassica* crop is likely to be affected, and their growing areas may be shifted to other regions conducive to the greater seed yield. Angadi et al. (2000) identified Indian mustard to have greater tolerance to heat and water stress than canola species. Niknam and Turner (1999) also reported that Indian mustard possesses several agronomic advantages over canola in relation to their response to heat stress.

The development of a rapeseed crop throughout the growing season may be characterized in terms of the timing of major phenological events such as seedling emergence, floral initiation, stem elongation, appearance of the first open flower, cessation of flowering, and maturity. Each developmental event affects the building of seed yield. In areas with short growing seasons (Canada and Southern Australia), where spring cultivars are normally grown, timing of flowering is crucial for seed development enabling it to complete before the onset of severe environmental stress (drought in Australia and frost in Canada). Flowering time is the major determinant to adapt these adverse environmental conditions (Thurling and Vijendra Das 1979). A rapid reduction in soil moisture significantly reduced the seed yield of *B. napus* in Western Australia. The seed yield of *Brassica napus* is largely determined by the temperature changes. The delay in sowing generally exposes a greater part of reproductive developmental phase to high-temperature stress. Simultaneously, the duration of the vegetative period before flowering decreased, and subsequently the dry matter production before flowering markedly reduced (Hodgson 1978). Campbell and Kondra (1977) demonstrated that early flowering cultivars of *Brassica* tend to be higher yielding than its late-flowering plants. These plants are higher yielding as a result of greater biomass production, more lateral branches and pods. The duration of pre-anthesis development was more important than post-flowering period in determining the yield of *Brassica* cultivars. The interaction between temperature and photoperiod helps in determining the time of flower initiation and for developing cultivars adapted to the changed temperature condition.

Quaderi et al. (2012) observed that higher temperature at seedling stage resulted in decreased stem height, leaf area, leaf area ratio, and water use efficiency, whereas it increased specific leaf weight, leaf weight ratio, and transpiration in canola (*Brassica napus*) plants. They have also demonstrated that higher temperature inhibited water stress-induced increases in ABA content. However, the water stress effect on canola seedling growth and dry matter production was greater than that of higher-temperature stress. Kutcher et al. (2010) through a chi square analysis study on the effect of temperature and precipitation in canola yield observed that the degree of reduction in yield was associated with higher temperature. The critical period for the impact of these factors on the yield determination is the beginning of July when canola plants are highly vulnerable to the changing climate. Canola producers may be able to modify the cultivation and management practices to mitigate the impact of these climatic factors on the yield on the basis of this study.

Iglesias et al. (2010) in a study on the response of broccoli plants to higher temperature observed that higher temperature up to 35 °C markedly reduced the P1P1 and P1P2 aquaporins in broccoli roots. The temperature at 40 °C and above adversely affected the membrane fluidity, which is determined by electrolyte leakage in the whole root. The aquaporins provide a signal mechanism for accumulation of heat stress, but water transport was mainly affected by the increased permeability through the lipid bilayer. Osmotic adjustment by altering membrane permeability is considered as an important mechanism of plant tolerance to heat stress.

The majority of crucifers are frost sensitive. Frost tolerance is an important factor in the successful cultivation of *Brassica* crops in many countries especially in India. Winter frosts in India are due to:

1. Dry cold waves enter across northwest frontier and move toward the northeast through northern Indian planes creating frost damage to *Brassica* crop in northwest India.
2. Radiation frost caused by a deficit in the radiation balance resulting into the loss of heat on the top of canopy making it the coldest point leading to frost damage.

Both types of frosts occur frequently on clear nights in January and the first fortnight of February. Frost damage occurs if the air temperature is below 10 °C at sunset (17.00 h) and the night is without clouds. The primary site of frost injury is the plasma membrane of the cell. Among the *Brassica* species, *B. campestris* varieties Span, Torch, and Bell are slightly susceptible to frost injury. *B. juncea* varieties "Varuna" and "RH-30" were highly susceptible, and *B. napus* variety "Norin-20" was most susceptible with 84.3 % seeds affected by frost (Banga et al. 1987). According to Dhawan (1985), the characters that were adversely affected by frost injury include seed yield, oil content, and harvest index; however, vegetative and reproductive phases were not very susceptible to frost damage. Plants growing at high altitude and in a wild state are likely to contain freezing resistance genes. While whole plants show little tolerance, individual cells may have greater variation. The selection at cell level may be a good option.

Suboptimal temperature during spring in Western Canada can delay seeding and reduce potential yield due to the short growing season. Acharya et al. (1983) in a study with *B. napus* and *B. campestris* observed the heritabilities for germination, and growth at 20 °C was close to zero in both species. However, the heritabilities of the same characters in populations grown at 10 °C ranged between 0.49 and 0.91 in *B. campestris* and between 0.23 and 0.64 in *B. napus*. The frost is the major abiotic factor limiting the productivity of *B. juncea* in North India. Richards and Thurling (1979) estimated the heritabilities of *B. napus* and *B. campestris* under high-temperature conditions. They observed that the heritability of yield was much higher in *B. napus* (0.44) than in *B. campestris* (0.10); however, the flowering time and 1000 seed weight had substantially higher heritability in *B. campestris*. They suggested that the selection for yield improvement under high-temperature condition can be made by using yield, pod numbers, and flowering time as selection criteria in *B. napus*. The most efficient index for *B. campestris* is the combination of yield, harvest index, 1000 seed weight, and seed number per pod. Biochemical characters such as leaf proline accumulation and leaf chlorophyll stability were also associated with high-temperature tolerance in *Brassica* (Richards 1978).

The improvement in *Brassica* crop yield is possible by developing the temperature tolerance in cultivars and through efficient management practices. The response to selection for environmental stress tolerance is unlikely to be very great because of low heritabilities observed for different stresses. An approach for temperature stress tolerance would be the use of pollen selection (Zamir 1983). It has been estimated that 60–70 % of genes function was similar during sporophytic and gametophytic phases of the life cycle. It has been shown that the selection pressure applied to pollens during germination and tube growth in vivo will increase the tolerance of progenies obtained through fertilization by the more vigorous pollen (Zamir et al. 1982). An investigation of this method with *B. napus* spring cultivars has revealed significant correlations between pollen tube growth at 10 °C and seedling growth under low temperature in the field and controlled environments (Thurling 1974).

4.6.1 Grain Quality

It was demonstrated that with the increase in temperature, both growth and maintenance respiration are increased and consequently lesser assimilates are available for seed biomass and that reduced the oil percentage in *Brassica* seeds. High temperature also adversely affected the pod and seed growth and duration of pod filling in *B. napus* by affecting their carbon balance (Whitfield 1992). Kawanabe (1979) demonstrated that the tropical species of oilseed *Brassica* show a very high percentage of saturated fatty acids. Crop grown in warm climate has a higher percentage of saturated fatty acids compared to the crop grown in the cooler region. Research in breeding and agronomy showed the enhancing trend of the energy flow through cropping systems. Plant breeders can exercise within the limits of available energy to elevate productivity, to combat the stresses, and to improve the seed quality (oil percentage and/or fatty acid composition). Yield level under high-temperature conditions can be enhanced by selecting genotypes of *Brassica* tolerant to high-temperature stress. The agronomic technologies such as timely irrigation and application of the higher level of fertilizers may also help in mitigating the higher-temperature stress. These cultivars should also have an efficient translocation of assimilates from sources to sinks.

Oilseed crop *Brassica* is generally grown under rain-fed conditions and is highly vulnerable to climate change. Walton et al. (1999) recorded that higher rainfall and cooler temperature produced higher yield and oil content (0.86 % increase for each 1 °C fall in daily mean temperature, 0.06 % increase for each 1 mm rain during seed development) in canola, whereas Hocking and Stapper (1993) reported an average 1.2–1.5 % oil loss per 1 °C rise in temperature. Temperature above 32 °C can result in substantial yield loss in *Brassica* species, which is dependent on the developmental stage of the crop (Morrison and Stewart 2002; Mendham and Salisbury 1995). Oil is deposited late during seed development and higher-temperature exposure at that time results in a significant reduction in oil content. Warmer condition also affects the fatty acid composition by favoring the production of saturated fatty acids, while cooler moist conditions favor the production of poly unsaturated fatty acids (Seiler 1983). The effect of higher temperature was significantly greater in protein content than that on oil concentration. Oil concentration shows an inverse relation with protein concentration with respect to temperature changes (Gunasekera et al. 2006). The fatty acid composition is also known to vary with higher temperature causing an increase in the percentage of monounsaturated fatty acid (oleic acid) and the decrease in polyunsaturated fatty acids (linolenic acid) in mustard (Yaniv et al. 1995; Triboi-Blondel and Renard 1999). Studies done by Anand and Nagrajan (2011) demonstrated that higher-temperature stress drastically reduced the yield by 54 % and 69 % in "Pusa Bold" and "Pusa Agrani" varieties of *B. juncea*, respectively. The number of pods/plant was reduced by 34 % and 48 % and the seed weight by 30 % and 40 % in these two cultivars. "Pusa Bold" showed a 23 % increase in seed weight/pod at brief period of stress without change in seed number, whereas in "Pusa Agrani" seed weight/pod was significantly less even under short period of high temperature. The increase in seed weight/pod in "Pusa Bold" resulted from lesser competition between pods and branches. A significant decrease in soluble starch and increase in crude protein content were observed at longer exposure to high temperature in Pusa Agrani. Oil content was increased by 3 % with more monosaturated and less polyunsaturated fatty acids under stress.

4.6.2 Adaptation and Mitigation

The agronomic strategies include short-term adjustments and long-term adaptations. Earlier planting in spring increases the length of growing season; thus, earlier planting using long-season cultivars will increase yield potential provided the moisture is adequate and the risk of heat stress is low. The earlier planting of a short-season cultivar would avoid the heat and water stress and will be the best option. Boomiraj et al. (2010) suggested adaptation strategies like late sowing, and growing longer-duration varieties would help in preventing high-temperature-induced yield losses of irrigated mustard crop in different locations. Similarly Faraji et al. (2009) observed that supplemental irrigation at reproductive stages, i.e., flowering and seed-filling periods, reduces the negative effects of high temperature in *Brassica napus* plants.

The long-term adaptations refer to major structural changes to overcome the adverse effect of high temperature. Parry et al. (1988a, b) reported that in central Europe, the area cultivated with winter wheat, maize, and vegetables increased, while the area of spring wheat, barley, and potato decreased. The following adaptation measures were identified:

1. Change in land allocation to stabilize the production. Crops with high variability in production need to be substituted by crops with lower productivity but more stable yield.
2. Crop substitution should also be looked in terms of conservation of soil moisture.
3. Crop breeding may be considered as another adaptation response to the change in temperature by the use of both traditional and biotechnological techniques to develop drought- and heat-resistant varieties.
4. Collection of genetic resources in germplasm banks to screen tolerances to heat and drought and for better compatibility to new agricultural technologies.
5. Genetic manipulations for possible rapid adaptations to warming-induced heat and drought stresses.
6. New land management techniques (minimum tillage, stubble mulching, etc.) or management strategies (e.g., irrigation rescheduling) may improve agricultural productivity under climate change stress conditions.
7. Intercropping, multi-cropping, and relay cropping may help in avoiding the high-/low-temperature stresses in crop plants.
8. Nutrient management such as fertilizer placement timing, changed crop rotation, and use of cover crops may help in mitigating the global warming stress effects in crop plants.

A number of agricultural management options including conservation tillage practices, crop residue management, cover crops, and altered crop rotations have been suggested as measures for climate warming stresses.

Crop yield in temperate regions is limited primarily by predictable or unpredictable environmental stresses. The predictable stresses are most severe toward the end of growing season. These stress effects can be circumvented by growing cultivars with a development pattern allowing for the avoidance of these stresses. However, a greater reduction in seed yield occurs if the stress is at the critical stages of development such as at the commencement of flowering. The most effective approach to meet this adverse stress effect is to breed for tolerance to the expected stresses.

References

Abrol YP, Uprety DC (1969) Effect of soil fertility levels on the protein quality of wheat grains. Indian Plant Physiol 12:211–215

Abrol YP, Uprety DC, Ram A (1971) Browning of Chapatis. J Food Sci Technol 8:26–27

Abrol YP, Uprety DC, Sharma A (1992) CO_2, Climate change and Agriculture. In: First Agricultural Science Congress, pp 63–70

Acevedo E (1991a) Effect of heat stress on wheat and possible selection tools for use in breeding for tolerance. In: Saunders DA (ed) Wheat for nontraditional warm areas. CIMMYI, Mexico, pp 401–421

Acevedo E (1991b) Improvement of winter cereal crops in Mediterranean environments: use of yield, morphological and physiological traits. In: Kohli MM (ed) Breeding for drought resistance in wheat, Top of Form Bottom of Form. Proceedings of the international symposium, Cordoba, Spain, 1987

Acharya SN, Dueck J, Downey RK (1983) Selection and heritabilities on canola/rapeseed for low temperature germination. Can J Plant Sci 63:377–384

Adachi K, Inoue J (1972) Effects of photoperiod, light intensity and components of culture medium on flower initiation in highly thermo sensitive paddy rice plants. Proc Crop Sci Soc Jpn 41: 78–82

Aggarawal PK, Sinha SK (1993) Effect of probable increase in carbon dioxide and temperature on productivity of wheat in India. J Agric Meteorol 48:811–814

Allen LH, Boote KJ (2000) Crop ecosystem responses to climate change: soybean. In: Reddy KR et al (eds) Climate change and global crop productivity. pp 133–160

Allen LH Jr, Albrecht SL, Colon-Guasp W et al (2003) Methane emission of rice increased by elevated carbon dioxide and temperature. J Environ Q 32:1978–1991

Anand A, NagrajanS (2011) Impact of climate change on quality of cereals and oilseed crops. In: Singh MP et al (eds) Climate change: impacts and adaptations of crop plants. Today and Tomorrows Printers and Publishers, Daryaganj, New Delhi, pp 33–53

Angadi SV, Cutforth HW, Miller PR et al (2000) Response of three brassica species to high temperature stress during reproductive growth. Can J Plant Sci 80:693–701

Anonymous (1991) Global climate change. Department of Environment, London, pp 1–16

Aronson EL, Mc Nulty SG (2009) Appropriate experimental ecosystem warming methods by ecosystem objectives and practicality. J Agric For Meteorol 64:1791–1799

Asana RD, Joseph CM (1964) Studies in physiological analysis of yield, VIl. Effect of temperature and light on the development of the grain of two varieties of wheat. Indian J Plant Physiol 7:86–101

Asana RD, Saini AD (1962) Studies in physiological analysis of yield, V. Grain development in wheat in relation to temperature, soil moisture and changes with age in the sugar content of the stem and in the photosynthetic surface. Indian J Plant Physiol 5:128–171

Asana RD, Williams RF (1965) The effect of temperature stress on grain development in wheat. Aust J Agric Res 16:1–13

Asatsuma S, Sawada C, Kitajima A, Asakura T, Mitsui T (2006) α-Amylase affects starch accumulation in rice grains. J Appl Glycosci 53:187–192

Ashida K, Iida S, Yasui T (2009) Morphological, physical, and chemical properties of grain and flour from chalky rice mutants. Cereal Chem 86(2):225–231

Ashilkari M, Sakakibara H, Lin S et al (2005) Cytokinin oxidase regulates rice grain production. Science 309(5735):741–745

Baker JM (2004) Yield responses of Southern US rice cultivars to CO_2 and temperature. Agric For Meteorol 122:129–137

Baker JT, Allen LH Jr (1993a) Contrasting crop species responses to CO_2 and temperature: rice, soybean and citrus. Vegetation 104(105):239–260

Baker JT, Allen LH Jr (1993b) Effects of CO_2 and temperature on rice. J Agric Meteorol 48(5):575–582

Baker JT, Allen LH Jr, Boote KJ et al (1989) Response of soybean to air temperature and carbon dioxide concentration. Crop Sci 29:98–105

Balla K, Bencze S, Janda T et al (2009) Analysis of heat stress tolerance in winter wheat. Acta Agron Hungar 57:437–444

Banga SK, Banga SS, Labana KS et al (1987) Screening for frost tolerance in Indian mustard (*Brassica juncea* L. Coss.). In: Proceedings 7th international rapeseed congress, Poland, pp 771–779

Barnabos B, Jager K, Feher A (2008) The effect of drought and heat stress on reproductive processes in cereals. Plant Cell Environ 31:11–18

Battisi DS, Naylor RL (2009) Historical warnings of future food insecurity with unprecedented seasonal heat. Science 323:240–244

Batts GR, Morison JIL, Ellis RH (1997) Effects of CO_2 and temperature on growth and yield of crops of winter wheat over several seasons. Eur J Agron 7:43–52

Bechoux N, Bernier G, Lejeune P (2000) Environmental effect on early stage of tassel morphogenesis in maize (*Zea mays*, L.). Plant Cell Environ 23:91–98

Bhabdwaj RBL (1978) New agronomic practices. In: Jaiswal PL (ed) Wheat research in India 1966–76. Indian Council of Agricultural Research, New Delhi, pp 79–98

Bhuller SS, Jenner CF (1985) Differential response to high temperatures of starch and nitrogen accumulation in the grains of four cultivars of wheat. Aust J Plant Physiol 12:363–375

Bhuller SS, Jenner CF (1986) Effect of temperature on the conversion of sucrose to starch in the developing wheat endosperm. Aust J Plant Physiol 13:605–615

Bluementhal CS, Barlow FRW, Wrigley CW (1993) Growth environment and wheat quality: the effects of heat stress on dough properties and gluten protein. J Cereal Sci 18:3–21

Blum A, Klueva N, Nguyen HT (2001) Wheat cellular thermo tolerance is related to yield under heat stress. Euphytica 117:117–123

Bonde MR, Nester SE, Berner DK (2012) Effect of daily temperature highs on development of Phakopsora pachyrhizi on soybean. Phytopathology 102:761–768

Boomiraj K, Chakrabarti B, Aggarawal PK et al (2010) Assessing the vulnerability of Indian mustard to climate change. Agric Ecosyst Environ 138:265–273

Brini F, Saibi W, Amaral I (2010) Wheat dehydrin dhn-5 exerts a heat protective effect on B glucosidase and glucose oxidase activities. Biosci Biotechnol Biochem 74:1050–1054

Cairns JE, Crossa J, Zaidi PH et al (2013) Identification of drought, heat and combined drought and heat tolerant donors in maize. Crop Sci 53:1335–1346

Campbell SA, Close TJ (1997) Dehydrins: genes, proteins, and associations with phenotypic traits. New Phytol 137(1):61–74

Campbell DC, Kondra ZP (1977) Growth pattern analysis of three rapeseed cultivars. Can J Plant Sci 57:707–712

Castro M, Peterson CJ, Rizza MD (2007) Influence of heat stress on wheat grain characteristics and protein molecular weight distribution. In: Buck HT et al (eds) Wheat production in stressed environment. Springer, pp 365–371

Chakravarty NVK, Shastri PSN (1983) Phenology and accumulated heat unit relationships in wheat under different planting dates in Delhi region. Agric Sci Prog 1:32–42

Chamura S, Honma M (1973) Injury of low temperature to transplanted young seedling of rice plant (in Japanese). Nigata Agric Sci 25:1–9

Chander S, Reji G, Aggarawal PK (2009) Assessing impact of climate change on rice gundhi bug using a population dynamics simulation model. In: Aggarwal PK (ed) Global climate change and Indian Agriculture, Case studies from the ICAR Network Project. ICAR, New Delhi, pp 37–38

Chaudhary TN, Ghildiyal BP (1970) Influence of submerged soil temperature regimes on growth, yield and nutrient composition of rice plant. Agron J 62:281–285

Chaudhary RS, Singh S, Nehra R (2005) Effect of replacement of maize by graded levels of pearl millet (*Pennisetum typhoides*) on broiler performance. Indian J Animal Sci 75(9):1092–1093

Cheng W, Hidemitsu S, Kazuyuki Y et al (2009) Interactions of elevated CO_2 and night temperature on rice growth and yield. Agric For Meteorol 149:51–58

Cicchino M, Rattalino Edreria JI, Otegui ME (2010) Heat stress during late vegetative growth of maize: effects on phenology & assessment of optimum temperature. Crop Sci 50:1431–1437

Coast O, Elis RH, Murdoch AJ et al (2015) High night temperature induces contrasting responses for spikelet fertility, spikelet tissue temperature, flowering characters and grain quality in rice. Funct Plant Biol 42:149–161

Conaty WC, Burko JJ, Mahan JR (2012) Determining the optimum plant temperature of cotton physiology and yield to improve plant based irrigation scheduling. Crop Sci 52:1828–1836

Cure JD, Acock B (1986) Crop responses to carbon dioxide doubling-a literature survey. Agric For Meteorol 38:127–145

Curry RB, Jones JW, Boote KJ et al (1995) Response of soybean to predicted climate change in the USA. In: Climate change and agriculture: analysis of potential international impacts, (climate change an), American Society of Agronomy, pp 163–182

Dale RF (1983) Temperature perturbations in Midwestern and Southeastern United States important for crop production. In: Raper CD et al (eds) Crop reactions to water and temperature stresses in humid and temperate climates. Westview Press, Colorado, pp 21–32

Daniel C, Triboi E (2000) Effects of temperature and nitrogen nutrition on the grain composition of winter wheat: effects on gliadin content and composition. J Cereal Sci 32:45–56

Dhawan AK (1985) Freezing in oilseed Brassica sps.: some factors affecting injury. J Agr Sci Camb 104:513–518

Dias AS, Bagulho AS, Lidon FC (2008) Ultrastructure and biochemical traits of bread and durum wheat grains under heat stress. Brazz J Plant Physiol 20:323–333

Dupont FM, Hurkman WJ, Vensel WH et al (2006) Protein accumulation and composition in wheat grains: effects of mineral nutrients and high temperature. Eur J Agron 25:96–107

Dupuis I, Dumas C (1990) Influence of temperature stress on in vitro fertilization and heat shock protein synthesis in maize (*Zea mays* L.) reproductive tissues. Plant Physiol 94:665–670

Easterling W, Aggarwal P, Batima P et al (2007) Food, fibre and forest products. In: Oarry ML et al (eds) Climate change 2007: impacts, adaptation and vulnerability. Cambridge University Press, Cambridge, pp 273–313

Egli DB (2004) Seed fill duration and yield of grain crops. Adv Agron 83:243–279

Enami I, Kitmura M, Tomo T (1994) Is the primary cause of thermal inactivation of oxygen evolution in spinach PSll, membranes release of the extrinsic 33kDa protein or of Mn? Biochem Biophys Acta 186:52–58

EPA (Environmental Protection Agency) (1989) The potential effects of global climate change on the United States. In: National studies vol 2. Review of the report to congress. US Environmental protection agency, Washington, DC, p 261

FAO (Food and Agriculture Organization) (2010) Crop production. URL: http://www.fao.org/docrep/017/i3138e/i3138e03.pdf. Accessed 10 Aug 2016

FAO (Food and Agriculture Organization) (2011) Food & Agriculture organization of United Nations. http://faostat3

FAOSTAT (2010) Food and Agriculture Organization statistical database. FAO, Rome

Faraji A, Latifi N, Soltani A et al (2009) Seed yield and water use efficiency of canola (Brassica napus) as affected by high temperature stress and supplemental irrigation. Agri c Water Manage 96:132–140

Fischer RA (1985) Number of kernels in wheat crops and the influence of solar radiation and temperature. CIMMYT Institutional Multimedia Publications Repository. J Agric Sci Camb 105:447–461

Fitzgerald MA, Mc Couch SR, Hall RD (2009) Not just a grain of rice: the quest for quality. Trends Plant Sci 14(3):133–139

Frei OM (2000) Changes in yield physiology of corn as a result of breeding in northern Europe. Maydica 45:173–183

Fuhrer J, Gregory PJ (2014) Climate change impact and adaptation in agricultural systems: soil ecosystem management in sustainable agriculture, vol 5. CABI, Boston

Gesch RW, Kang IH, Gallo-Meagher M (2003) Rubisco expression in rice leaves is related to genotypic variation of photosynthesis under elevated growth CO_2 and temperature. Plant Cell Environ 26(12):1941–1950

Ghosh S, Gepstein S, Glick BR (1989) Thermal regulation of phosphoenolpyruvate carboxylase and ribulose-1,5-bisphosphate carboxylase in C3 and C4 plants native to hot and temperate climates. Plant Physiol 90:1298–1304

Ghosh M, Mandal BK, Mandal BB et al (2004) The effect of planting date and nitrogen management on yield and quality of aromatic rice (*Oryza sativa*). J Agric Sci 142:183–191

Gibson LR, Mullen RE (1996) Soybean seed composition under high day and night growth temperatures. J Am Oil Chem Soc 73(6):733–737

Gifford RM (1995) Whole plant respiration and photosynthesis of wheat under increased CO_2 concentration and temperature: Long term vs short term distinctions for modeling. Glob Change Biol 1:385–396

GlaubitzU LX, Kohl KI et al (2014) Differential physiological responses of different rice (*Oryza sativa*) cultivars to elevated night temperature during vegetative growth. Funct Plant Biol 41:437–448

Gunasekera CP, Martin LD, Siddique KHM et al (2006) Genotype by environment interactions of Indian mustard (*Brassica juncea* L.) and canola (*Brassica napus* L.) in Mediterranean-type environments: II. Oil and protein concentrations in seed. Eur J Agron 25(1):13–21

Guo WJ, Lee N (2006) Effect of leaf and plant age, and day/night temperature on net CO_2 uptake in Phalaenopsis amabilis var. formosa. J Am Soc Hortic Sci 131:320–326

Hamid A, Haque M, Khanam M et al (2003) Photosynthesis, growth and productivity of rice under elevated CO_2. Indian J Plant Physiol 131:253–258

Harris K, Subudhi PK, Borrell A et al (2007) Sorghum stay green QTL individually reduce post flowering drought- induced leaf senescence. J Expt Bot 58:327–338

Hasegawa T, Shimono H, Fujimura S et al (1999) Effects of root cooling on growth and water status in rice. Rep. Hokkaido Branch. Crop Sci Soc Jpn 40:41–42

Hasegawa T, Fukuoka M, Yoshimoto M et al (2006) Increased canopy temperature due to retarded transpiration induced spikelet sterility in rice. Jpn J Crop Sci 75:226–227

Hawker JS, Jenner CF (1993) High temperature effects the activity of enzymes in the committed pathway of starch synthesis in developing wheat endosperm. Aust J Plant Physiol 20:197–209

He GC, Kogure K, Suzuki H (1990) Development of endosperm and synthesis of starch in rice grain. Lll. Starch property affected by temperature during grain development. Jpn J Crop Sci 59:340–345

Hebbar KB, Venugopalan MV, Prakash AH et al (2013) Simulating the impacts of climate change on cotton production in India. Climate Change 118:701–713

Heinemann JA, Rosen H, Savill M et al (2006) Environmental arrays: a possible approach of predicting changes in waterborne bacterial disease potential. Environ Sci Tech 40:7150–7156

Herzog H (1986) Source and sink during the reproductive period of wheat. Adv Agron Crop Sci S8:1–104

Hocking PJ, StapperM (1993) Effect of sowing time and nitrogen fertilizer rate on the growth, yield and nitrogen accumulation of canola, mustard and wheat. In: Proceedings of the 9th Aus. Research assembly on Brassica Wagga, NSW, New South Wales agriculture, Wagga NSW, pp 33–46

Hodges HF, Reddy KR, McKinion JM et al (1993) Temperature effects on cotton. Missipi Agricultural and Experimental Station. Bulletin no. 990, Missipi State University, p 15

Hodgson AS (1978) Rapeseed adaptations northern New South Wales. 11. Predicting plant development of Brassica campestris L. and Brassica napus L. and its implications on planting time, designed to avoid water deficit and frost. Aust J Agric Res 29:711–726

Hofstra G, Hasketh JD (1975) The effects of temperature and CO_2 enrichment on photosynthesis in soybean. In: Marcelle R, Junk W (eds) Environmental and biological control of photosynthesis. Publication, The Hague, pp 71–80

Horie T (1993) Predicting the effect of climate variation and elevated CO_2 on rice yield in Japan. J Agric Meteorol 48:567–574

Horie T, Baker JT, Nakagawa H et al (2000) Crop ecosystem responses to climate change. In: Reddy KR, Hidges HF (eds) Rice, climate change and global productivity. Cambridge University Press, Cambridge, pp 81–106

Horie T, LubisI,Takai T et al (2003) Physiological traits associated with high yield potential in rice. In: Mew TW et al (eds) Rice science: innovations and impact for livelihood. Los Banos (Philippines), IRRI, Philippines, pp 117–145

Horie T, Costa A, Kim TH et al (2007) OsHKT2:I transporter mediates large Na^+ influx component in to K^+-starved roots for growth. EMBO J 26:3003–3014

Howard A (1924) Crop production in India. Oxford University Press, London

Huang M, Guo Z (2005) Responses of antioxidant system to chilling stress in two rice cultivars differing in sensitivity. Biol Planta 49:81–84

Huang M, Zhang W, Jiang L et al (2013a) Impact of temperature changes on early rice productivity in a sub-tropical environment of China. Field Crop Res 146:10–15

Huang M, Jiang L, Zou Y et al (2013b) On farm assessment of effect of low temperature at seedling stage on early season rice quality. Field Crop Res 141:63–68

Hungria M, Vargas MA (2000) Environmental factors affecting N_2 fixation in grain legumes in the tropics, with an emphasis on Brazil. Field Crop Res 65(2):151–164

Iglesias-Acosta M, Martínez-Ballesta MC, Teruel JA et al (2010) The response of broccoli plants to high temperature and possible role of root aquaporins. Environ Exp Bot 68(1):83–90

Imai K, Okamota- Sato M (1991) Effects of temperature on CO_2 dependence of gas exchanges in C3 and C4 crop plants. J Crop Sci Jpn 54:413–418

Imai K, Colman DF, Yanagisawa T (1985) Increase of atmospheric partial pressure of carbon dioxide and growth and yield of rice (*Oryza sativa* L.). Jpn J Crop Sci 54:413–418

Imaki T, Tokunaga S, Obara S (1987) High temperature induced spikelet sterility of rice in relation to flowering time. Jpn J Crop Sci 56:209–210

Ingram KT, Manalo PA, Namuco OS et al (1995) Interactive effects of elevated carbon dioxide and temperature on rice growth and development. In: Peng et al (eds) Climate change and rice, Springer, Berlin, pp 278–287

IPCC (International Panel on Climate Change) (2007) Summary for policy makers. In: Parry ML et al (eds) Impacts, adaptation and vulnerability, contribution of working group II to the fourth assessment report of the Intergovernmental Panel on Climate Change. Cambridge University Press, Cambridge

IPCC (International Panel on Climate Change) et al (2013) In: Stocker TF, Platner GK, Tignor M (eds) The physical science basis. Contribution of working group I to the fifth assessment report of IPCC. Cambridge University Press, Cambridge

Jenner CF (1991) Effects of exposure of wheat ears to high temperature on dry matter accumulation and carbohydrate metabolism in the grain of two cultivars. II. Carry over effects. Aust J Plant Physiol 18:179–190

Jenner CF, Denyer K, Guerin J (1995) Thermal characteristics of soluble starch synthase from wheat endosperm. Aust J Plant Physiol 22:703–709

Jiang C, Zeng ZB (1995) Multiple trait analysis of genetic mapping for quantitative trait loci. Genetics 140:1111–1127

Jones P, Allen LH Jr, Jones JW et al (1985) Photosynthesis and transpiration responses of soybean canopies to short and long term CO_2 treatments. Agron J 77:119–126

Kaukoranta T (1996) Impact of global warming on potato late blight: risk, yield loss and control. Agric Food Sci Finl 5:311–327

Kawanabe S (1979) Fatty acid composition and iodine values of seeds of oil and fat crops in relation to climatic conditions. J Trop Agric 23:11–20

Khush GS, Poule CM, Dela Cruz N (1979) Rice grain quality evaluation and improvement at IRRI. In: Chemical aspects pf rice grain quality, IRRI, Philippines, pp 21–31

Kim HY, Horie T, Nakagawa H (1996) Effects of elevated CO_2 concentration and high temperature on growth and yield of rice. Jpn J Crop Sci 65:644–651

Kim J, Shon J, Lee CK et al (2011) Relationship between grain filling duration and leaf senescence of temperate rice under high temperature. Field Crop Res 122:207–213

Kim SH, Yang Y, Timlin D et al (2012) Modeling temperature responses of leaf growth, development& biomass in maize with MAIZSIM. Agron J 104:1523–1537

Kimball BA, Pinter PJ Jr, Mauny JR (1992a) Cotton leaves and Boll temperature in the 1989 FACE experiment. CRC Cr Rev Plant Sci 11:233–240

Kimball BA, Pinter PJ Jr, Mauny JR (1992b) Cotton leaves and Boll temperature in the 1989 FACE experiment. CRC Cr Rev Plant Sci 11:233–240

Kingsolver JG (1989) Weather and the population dynamics of insects: integrating physiological and population ecology. Physiol Zool 62:314–334

Krishnan P, Rao AVS (2005) Effects of genotype and environment on seed yield and quality of rice. J Agric Sci 143:283–292

Krishnan P, Swain DK, Basker C et al (2007) Impact of elevated CO_2 and temperature on rice yield and methods of adaptation as evaluated by crop simulation studies. Agric Ecosyst Environ 122(2):233–242

Kucharik CJ, Serbin SP (2008) Impacts of recent climate change on Wisconsin corn and soybean yield trends. Environ Res Let. doi:10.1088/1T48-9326/3/3/034003

Kureck I, Chang TK, Bertain et al (2007) Enhanced thermostability of Arabidopsis Rubisco activase improves photosynthesis and growth rates under moderate heat stress. Plant Cell 19:3230–3241

Kutcher HR, Warland JS, Brandt SA (2010) Temperature and precipitation effects on canola yields in Saskatchewan, Canada. Agric Forest Meteorol 150:161–165

Lafitte HR, Banziger M, Edmeades GO (1997) Maize population. Improvement for low soil nitrogen. In: Edneades GO, Banziger M, Mickelson HR et al (eds) Developing drought and low nitrogen tolerant maize. Proceedings of symposium, 25–291996 CIMMYT, El Baten, Mexico

Lal M, Singh KK, Rathore LS et al (1998) Vulnerability of rice and wheat yields in NW India to future changes in climate. Agric Forest Meteorol 89:101–114

Lawlor DW, Key AJ (1993) Understanding photosynthetic adaptation to changing climate. In: Fowden L, Mansfield TA et al (eds) Plant adaptations to environmental stress. Chapman and Hall, London, pp 85–106

Lawlor DW, Mitchell RAC (2000) Crop ecosystem responses to climate change: wheat. In: Reddy KR et al (eds) Climate change and global crop productivity, pp 57–80

Lea PJ, Leegood RC (1999) Plant biochemistry and molecular biology. Wiley, Chichester

Leach JE, Darby RJ, Williams IH, Fitt BDL et al (1994) Factors affecting growth and yield of winter oilseed rape (*Brassica napus*),1985-1989. J Agric Sci 122:405–413

Leakey ADB, Xu F, Gillespie KM et al (2009) Genomic basis for stimulated respiration by plants growing under elevated carbon dioxide. Proc Natl Acad Sci 106:3597–3602

Lee KH, Akita S (1999) Variation in temperature coefficient of dark respiration in rice. In: Horie T et al (eds) World food security and crop production technologies for tomorrow, Proceedings of international symposium, Crop Science Society of Japan, pp 282–283

Leegood RC (2002) C4 photosynthesis: principles of CO_2 concentration and prospects for its introduction in to C3 plants. J Exp Bot 3:581–590
Leemans R, Solomon AM (1993) Modeling the potential change in yield and distribution of the earth's crops under a warmed climate. Climate Res 3:79–96
Lin W, Ziska LH, Namuco OS (1997) The interaction of high temperature and elevated CO_2 on photosynthetic acclimation of single leaves of rice in situ. Physiol Plant 99(1):178–184
Lin Q, Liu Z, Xiao H et al (2009) Studies on the pasting and rheology of rice starch with different protein residual. In: Computer and computing technologies in agriculture III. Springer, Berlin/Heidelberg, pp 407–419
Liu TS (1997) Quaternary environment. Science Press, Beijing
Liu Y, Yan J et al (2011) Maize leaf temperature responses to drought: thermal imaging and quantitative loci (QTL) mapping. Environ Exp Bot 71:158–165
Lobell DB, Asner GP (2003) Climate and management contributions to recent trends in US agricultural yields. Science 299:1032
Lobell DB, Burkee MB (2010) On the use of statistical models to predict crop yield responses to climate change. Agric For Meteorol 150:1443–1452
Lobell DB, Oritz-Monasterio IJ, Asner GP et al (2005) Analysis of wheat yield and climatic trends in Mexico. Field Crop Res 94:250–256
Lobell DB, Banziger M, Magorokosho C (2011) Nonlinear heat effects on African maize as evidenced by historical yield trials. Nat Clim Change 1:42–45
Loka DA, Oosterhuis DM (2010) Effect of high night temperatures on cotton respiration, ATP levels and carbohydrate content. Environ Exp Bot 68:258–263
Lonenko IF, Anisimov AV, Dautova VR (2010) Effect of temperature on water transport through aquaporin. Biol Planta 54:488–494
Long SP (1991) Modification of the response of photosynthetic productivity to rising temperature by atmospheric CO_2 concentrations: has its importance been underestimated? Plant Cell Environ 14:729–739
Long SP, Ort DR (2010) More than taking the heat: crops and global change. Curr Opin Plant Biol 13:1–8
Long SP, Ainsworth EA, Rogers A et al (2004) Rising atmospheric carbon dioxide: plants face the future. Ann Rev Plant Biol 55:591–628
Lu Z, Zeiger E (1994) Selection of higher yield and heat resistance in pima cotton has caused genetically determined changes in stomatal conductance. Physiol Plant 92:273–278
Lu D, Shen X, Cai X et al (2014) Effects of heat stress during grain filling on the structure and thermal properties of waxy maize starch. Food Chem 143:313–318
Lukatin AS, Naraikina NV (2011) Assessing the effect of temperature stressors on maize plants from the change in antioxidant activity. Russian Agric Sci 37:364–366
Luo Y, Teng PS, Fabellar NG (1997) A rice leaf blast combined model for simulation of epidemics and yield loss. Agric Syst 53:27–39
Mackill DJ, Ismail AM, Pamplona AM et al (2010) Stress tolerant rice varieties for adaptation to a changing climate. Crop Environ Bioinfo 7:250–259
Maiorano A, Mancini MC (2010) Water relationships & temperature interactions in maize grain during maturation. Field Crop Res 119:304–307
Majoul T, Bancel E, Triboi RE et al (2003) Proteomic analysis of the effect of heat stress on hexaploid wheat grains. Characterization of heat responsive proteins from total endosperm. Proteomics 3:175–183
Mann RF (1987) Basmati rice: wonder of Pakistan's agriculture. IRRI News Lett 36:23–28
Matsui T, Namuco OS, Ziska LH et al (1997) Effect of high temperature and CO_2 concentration on spikelet sterility in Indica rice. Field Crop Res 51:213–221
Mattson WJ, Haack RA (1987) The role of drought in outbreaks of plant eating insects. Bioscience 37:110–118
Mc Gregor DI (1981) Pattern of flower and pod development in rapeseed. Can J Plant Sci 61: 275–282

Mc Martin W, Frank AB, Heintz RH (1974) Economics of shelterbelt influence on wheat yields in North Dakota. J Soil Water Conserv 29:87–91

Mendham NJ, Salisbury PA (1995) Physiology; crop development, growth and yield. In: Kimber D et al (eds) Brassica oilseeds: production and utilization. CAB International, Slough, pp 11–64

Miedema P (1982) The effects of low temperature on Zea mays. Adv Agron 35:93–128

Miller JD (2001) Factors that affect the occurrence of fumonisin. Environ Health Perspect 109:321–324

Mitchell RAC, Mitchell VJ, Driscoll SP et al (1993) Effects of increased CO_2 concentration and temperature on growth and yield of winter wheat at 2 levels of nitrogen application. Plant Cell Environ 16:521–529

Mitchell JFB, Johns TC, Gregory JM et al (1995) Climate response to increasing levels of greenhouse gases and sulphate aerosols. Nature 376:501–504

Mohammad AR, Taropley L (2009) High night time temperatures affect rice productivity through altered pollen germination and spikelet fertility. Agric For Meteorol 149:999–1008

Mohammad AR, Taroply L (2011) Characterization of rice (*Oryza sativa*): physiological responses to alpha tocopherol, glycine betaine and salicylic acid application. J Agric Sci 3:3–13

Mohapatra PK, Patel R, Sahu SK (1993) Time of flowering affects grain quality and spikelet partitioning within the rice panicle. Aust J Plant Physiol 20:231–241

Morita S, Yonemaru JI, Takanashi JI (2005) Grain growth and endosperm cell size under high night temperature in rice. Ann Bot 95:695–701

Morrison JM, Stewart DW (2002) Heat stress during flowering in summer Brassica. Can J Bot 42:797–803

Muchow R (1990) Effect of high temperature on the rate and duration of grain growth in field grown *Sorghum bicolor* (L) Moench. Aust J Agric Res 41:329–337

Muller S, Vensel WH, Kasarda DD (1998) Disulphide bonds of adjacent cysteine reduces in molecular weight subunits of wheat glutenin. J Cereal Sci 27:109–116

Nagai T, Makino A (2009) Differences between rice and wheat in temperature responses of photosynthesis and plant growth. Plant Cell Physiol 50:744–755

Nagrajan S, Jagdish SVK, Hariprasad AS et al (2010) Effect of night temperature and radiation on growth, yield and grain quality of aromatic and non-aromatic rice. Agric Ecosyst Environ 138:245–250

Nakagawa H, Horie T, Matsui T (2003) Effect of climate change on rice production and adaptive technologies. In: Mew TW, Brar DS, Dawe D et al (eds) Rice science: innovations and impact for livelihood. International Rice Research Institute, Philippines, pp 635–658

Narayanan S, Prasad PVV, Fritz AK et al (2015) Impact of high night time and high day time temperature stress on winter wheat. J Agri Crop Sci 201:206–218

Nigam PK (1977) Physiology of tillering in wheat. Ph.D. thesis, Indian Agricultural Research Institute, New Delhi, India

Niknam SR, Turner DW 1999) A single drought event, at different stages of development, has different effects on the final yield of Brassica napus cv. Monty and B. juncea line 397-23-2-3–3. In: Shea G (ed) 1999 Oilseed crop updates agriculture, Western Australia, Northam, pp 14–15

NishiyamaI (1984) Climatic influence on pollen formation and fertilization. In: Tsunoda S et al (eds) Jpn Sci Soc PressTokyo, pp 153–171

NishiyamaI, Blanco L (1980) Avoidance of high temperature sterility by flower opening in the early morning. Jpn AgricRese Q 14:116–117

O'Byrne DMJ (1995) The effects of a low fat diet high in monounsaturated fatty acids on serum lipids, apolipoproteins, and lipoproteins in postmenopausal women with hypercholesteremia. Dissertation, University of Florida

Oh-e I, Saitoh K, Kuroda T (2007) Effects of high temperature on growth, yield and dry-matter production of rice grown in the paddy field. Plant Prod Sci 10:412–422

Ohnishi S, Miyoshi T, Shirai S (2010) Low temperature stress at different flower development stages affects pollen development, pollination and pod set in soybean. Environ Expt Bot 69: 56–62

Ortiz R, Sayre KD, Govarts B et al (2008) Climate change: can wheat beat the heat? Agric Ecosyst Environ 126:46–58
Osada A, Saciplapa V, Rahong M et al (1973) Abnormal occurrence of empty grains of indica rice plants in the dry, hot season in Thailand. Proc Crop Sci Soc Jpn 42:103–109
Ottman MJ, Kimball BA, White JW et al (2012) Wheat growth response to increased temperature from varied planting dates and supplemental Infra red heating. Agron J 104:7–16
Pan D (1996) Soybean responses to elevated temperature and doubled CO_2. Dissertation, University of Florida
Pandey VK, Sharma KC (1996) Crop diversification and self-sufficiency in food grains. Indian J Agric Econ 51(4):644–651
Parry ML, Carter TR, Konijin NT (1988a) The impact of climate variation in agriculture, vol 1. Assessment in cool temperate and cold regions. Kluwer, Dordrecht
Parry ML, Konijn NT, Carter TR (1988b) The impact of climate variations on agriculture, vol 2, Assessments in semi-arid regions. Kluwer Academy Publishers, Dordrecht, p 764
Parry ML, Rosenzweig C, Iglesias A et al (2004) Effects of climate change on global food production under SRES emissions and socio-economic scenarios. Glob Environ Change 14(1):53–67
Pasquin EM, Lafarge T, Tubana B (2008a) Transplanting young seedlings in irrigated rice fields: Early and high tiller production enhanced grain yield. Field Crop Res 105:141–155
Pasquin EM, Eberbach P, Lafarge T et al (2008b) Dry matter production and partitioning in rice in response to elevated temperature, Global Issues Paddock Action. In: Unkovich MJ (ed) Proceedings of the 14th Australian Agronomy conference, September, 2008, Australian Society of Agronomy, Adelaide, South Australia
Patil RH, Laegdsmand M, Oleson JE et al (2010) Growth and yield response of winter wheat to soil warming and rainfall patterns. J Agric Sci 148:533–566
Peng S, Huang J, Sheehy JE et al (2004) Rice yields decline with higher night temperatures from global warming. Proc Natl Acad Sci 101:9971–9975
Penning de veris FWT, Jenson DM, Ten Berge HFM et al (1989) Simulation of ecophysiological processes in several annual crops In: Simulation monograph, PUDOC, Wageningen, p 271
Phadke KG, Ghai S (1994) Effect of global warming on insect populations and crop damage. Shashpa 1:275–280
Pipe EL, Boote KJ (1999) Temperature and cultivar effects on soybean seed oil and protein concentrations. J Am Oil Chem Soc 76:1233–1241
Prasad R (1989) Response of Brassica species to the thermal environment under Delhi conditions. Dissertation, Indian Agricultural Research Institute, New Delhi
Prasad PVV, Djanaguiraman M (2014) Response of floret fertility and individual grain weight of wheat to high temperature stress: sensitive stages and thresholds for temperature and duration. Funct Plant Biol 41(12):1261–1269
Prasad PVV, Boote KJ, Allen LH Jr (2002) Effects of elevated temperature and carbon dioxide on seed set and yield of kidney bean (*Phaseolus vulgaris* L.). Glob Change Biol 8:710–721
Prasad PVV, Allen LH, Boote KJ (2005) Crop responses to elevated carbon dioxide and interaction with temperature: grain legumes. J Crop Improv 12:113–155
Prasad PVV, Boote KJ, Allen LH Jr (2006) Species, ecotype and cultivar differences in spikelet fertility and harvest index of rice in response to high temperature stress. Field Crop Res 95(2):398–411
Prasad PVV, Staggenborg SA, Ristic Z (2008) Impacts of drought and/or heat stress on physiological, developmental, growth and yield processes of crop plants. http://www-personal.ksu.edu/-vara/Prasad-pvv-aasm-chapter%2011.pdf. Accessed on 9 Mar 2016
Putch AB, Thuzar M, Mondal MMA et al (2013) Soybean (*Glycine max* L. Merrill) seed yield response to high temperature stress during reproductive growth stages. Aust J Crop Sci 7: 1472–1479
Qaderi M, Kurepin VL, Reid MD (2012) Effects of temperature and watering regime on growth, gas exchange and abscisic acid content of canola (*Brassica napus*) seedlings. Environ Exp Bot 75:107–113

Rahman M (2003) Optimum sowing time for higher seed yield and quality in rice. Seed Res 31:137–140

Randall PJ, Moss HJ (1990) Some effects of temperature regime during grain filling on wheat quality. Aust J Agric Res 41:603–617

Rang ZW, Jagdish SVK, Zhou QM et al (2011) Effect of high temperature and water stress on pollen germination and spikelet fertility in rice. Environ Exp Bot 70:58–65

Ravindra V (1985) Studies on photosynthetic efficiency, biomass production and distribution as a function of temperature and photosynthetically active radiation in *Brassica juncea* L.Pusa Bold, Dissertation, Indian Agricultural Research Institute, New Delhi

Rawson HM (1988) Effects of high temperatures on the development and yield of wheat and practices to reduce deleterious effects. In: Klatt R (ed) Wheat production constraints in tropical environments. CIMMYT, Mexico, pp 44–63

Rawson HM (1992) Plant responses to temperature under conditions of elevated CO_2. Aust J Bot 40:473–490

Rawson HM, Bagga AK (1979) Influence of temperature between floral initiation and flag leaf emergence on grain number in wheat. Aust J Plant Physiol 6:391–400

Rebetzke GJ, Condoa AG, Richards RA (1996) Rapid screening of leaf conductance in segregating wheat populations. In: Richards RA et al (eds) Proceedings of eighth assembly, Wheat Breeding Society of Australia, Australia, pp 130–134

Reddy VR, Reddy KR, Baker DN (1991a) Temperature effects on growth and development of cotton during the fruiting period. Agron J 83:211–217

Reddy VR, Baker DN, Hodges HF (1991b) Temperature effects on cotton canopy growth, photosynthesis and respiration. Agron J 83:699–704

Reddy KR, Reddy VR, Hodges HF (1992) Temperature effects on early season cotton growth and development. Agron J 84:229–237

Reddy KR, Hodges HF, McKinion JM (1997) A comparison of scenarios for the effect of global climate change on cotton growth and yield. Aust J Plant Physiol 24:707–713

Reddy KR, Robana RR, Hodges HF (1998) Influence of atmospheric CO_2 and temperature on cotton growth and leaf characteristics. Environ Exp Bot 39:117–129

Reddy VR, Pachepsky LB, Whisler FD (2002) Effect of temperature and photoperiod on development rates of nine soybean varieties in the Mississippi valley. Acta Hortic 593:201–207

Richards GA (1978) Seasonal and environmental variations in sediment accretion in Long Island salt marsh. Estuaries 1(10):29–35

Richards RA, Thurling N (1979) Genetic analysis of drought stress response in rapeseed (*Brassica campestris* and *B. napus*): Yield improvement and application of selection indices. Euphytica 28:169–177

Riley GJP (1981) Effect of high temperature on the germination of maize (*Zea mays* L.). Planta 151:68–74

Rosegrant MW, Msangi S, Ringler C et al (2008) International model for policy analysis of agricultural commodities and trade (IMPACT): model description, International Food policy Research Institute, Washington, DC. http://www.ifpri.org/themes/impact/impactwater.pdf Accessed 9 Mar 2016

Roth GW, Yocum JO (1997) Use of hybrid growing degree day rating for corn in the northeastern USA. J Prod Agric 10:283–288

Rowhani P, Lobell DB, Linderman M et al (2011) Climate variability and crop production in Tanzania. Af For Meteorol 151:449–460

Rowland-Bamford AJ, Baker JT, Allen LH Jr et al (1996) Interactions of CO_2 enrichment and temperature on carbohydrate accumulation and partitioning in rice. Environ Exp Bot 36: 111–124

Russell G, Wilson GW (1994) An agro-pedo-climatological knowledge-base of wheat in Europe. European Commission, Luxembourg, p 160

Sadok W, Sinclair TR (2010) Transpiration response of slow wilting and commercial soybean (*Glycine max* L.) genotypes to three aquaporin inhibitors. J Expt Bot 61:821–829

Sahoo KM, Saraswat VN (1988) Magnitude of losses in the yields of major crops due to weed competition in India. Pesticide Inform 14:2–9

Saini JS (1984) Agronomic considerations for raising oilseed production under irrigated conditions In: Research and developmental strategies for oilseed production in India. ICAR, New Delhi, pp 155–165

Sairam RK, Srivastava GC, Saxena DC (2000) Increased anti-oxidant activity under elevated temperature: a mechanism of heat stress tolerance in wheat genotypes. Biol Planta 43:245–251

Salem MA, Kakani VJ, Koti S et al (2007) Pollen based screening of soybean genotypes for high temperatures. Crop Sci 47:219–231

Salvucci ME, Crafts Brandner SJ (2004) Inhibition of photosynthesis by heat stress: the activation state of Rubisco as a limiting factor in photosynthesis. Physiol Planta 120:179–186

Salvucci ME, Osteryoung KW, Crafts Brandner SJ et al (2001) Exceptional sensitivity of Rubisco activase to thermal denaturation in vitro and in vivo. Plant Physiol 127:1053–1064

Sastry AB, Kumar A (1981) Variations in yield, its attributes and quality of Indian mustard in relation to planting time and levels of plant population. Indian J Agric Sci 51:27–32

Satake T (1995) High temperature injury. In: MatsuoT et al (eds) Science of the rice plant, vol 2, physiology, Food and Agriculture Policy Research Center, Tokyo, pp 805–812

Satake T, Hayashi H (1970) Male sterility caused by cooling treatment at the young microspore stage in rice plants: estimation of pollen development stage and the most sensitive stage to coolness. Proc Crop Sci Soc Jpn 39:468–473

Satake T, Yoshida S (1978) High temperature induced sterility in Indica rice at flowering. Jpn J Crop Sci 47:6–10

Savin R, Slafer GA, Galderini DF et al (1999) Final grain weight in wheat as affected by short periods of high temperature during pre- and postanthesis under field conditions. Aust J Plant Physiol 26:453–458

Schoper JB, Lambert RJ, Vasilas BL (1987a) Pollen viability, pollen shedding and combining ability for tassel heat tolerance in maize. Crop Sci 27:27–31

Schoper JB, Lambert RB, Vasilas BL et al (1987b) Plant factors controlling seed set in maize: the influence of silk, pollen, and ear, leaf water status and tassel heat treatment at pollination. Plant Physiol 83:121–125

Scofield GN, Hirosi T, Gaudron JA et al (2002) Antisense suppression of the rice sucrose transport gene. OsSUT1 leads to impaired grain filling and germination but does not affect photosynthesis. Funct Plant Biol 29:815–826

Seiler GJ (1983) Effect of genotype, flowering date and environment on oil content and oil quality of wild sunflower seeds. Crop Sci 23:1063–1068

Sharma KD, Kumar A (2010) Genotypic variation for agro physiological traits and their utilization as screening indices for drought tolerance in wheat. Indian J Genet 70(1):1–5

Sharma KD, Pannu RK, Tyagi PK et al (2003) Effect of moisture stress on plant water relations and yields of different wheat genotypes. Indian J Plant Physiol 8(10):99–102

Shaykewich CF (1995) An appraisal of cereal crop phenology modeling. Can J Plant Sci 72:1157–1162

Sheehy JE, Mitchell PL, Ferrer AB (2006) Decline in rice grain yields with temperature: models and correlations can give different estimates. Field Crop Res 98(2):151–156

Siddiqui BS, Rasheed M, Ghiasuddin FA et al (2000) Two insecticidal Tetranortriterpenoids from *Azadirachta indica*. Phytochemistry 53:371

Singh SK, Kakani VG, Braid D et al (2008) Assessment of cold and heat tolerance of winter grown canola (*Brassica napus* L.) cultivars by pollen based parameters. J Agron Crop Sci 194:225–236

Singh B, Kunze B, Satyanarayana T (2011) Developments in biochemical aspects and biotechnological applications of microbial phytases. Biotechnol Mol Biol Rev 6:69–87

Singh VV, Ram BB, Singh M et al (2014) Generation mean analysis for water stress tolerance parameters in Indian mustard (B. juncea L.Cern & Coss) crosses. SABRAO J Breed Gent 46: 76–80

Singletary GW, Banisadr R, Keeling PL (1994) Heat stress during grain filling in maize:effects on carbohydrate storage and metabolism. Aust J Plant Physiol 21:829–841

Sionit N, Strain BR, Flint EP (1987) Interactions of temperature and CO_2 enrichment on soybean: growth and dry matter partitioning. Can J Plant Sci 67:59–67

Slafer GA, Rawson HM (1994) Sensitivity of the wheat phasic development to major environmental factors: a re-examination of some assumptions made by physiologists and modellers. Aust J Plant Physiol 21:393–426

Slafer GA, Rawson HM (1995) Base and optimum temperatures vary with genotype and stage of development in wheat. Plant Cell Environ 18:671–679

Spiertz JHJ (1997) The influence of temperature and light intensity on grain growth in relation to the carbohydrate and nitrogen economy of the wheat plants. Netherland J Agric Sci 25:182–197

Stone PJ (2001) The effects of heat stress on cereal yield and quality. In: Basra AS (ed) Crop responses and adaptations to temperature stress. Food Products Press, Inghamton, pp 243–291

Sujatha KB, Uprety DC, Nageswara Rao D et al (2008) Up regulation of photosynthesis and sucrose-P synthase in rice under elevated carbon dioxide and temperature conditions. Plant Soil Environ 54:155–162

Swaminathan MS (1984) Rice. Sci Am 250:81–93

Tacarindua CR, Shiraiwa T, Homma K et al (2013) The effects of increased temperature on crop growth and yield of soybean grown in a temperature gradient chamber. Field Crop Res 154: 74–81

Tahir ISA, Nakata N (2005) Remobilization of nitrogen and carbohydrate from stems of bread wheat in response to heat stress during grain filling. J Agron Crop Sci 191:106–115

Tao F, Zhang Z (2011) Impact of climate change as a function of global mean temperature: maize productivity and water use in China. Climate Change 105:409–432

Tashiro T, Wardlaw IF (1990) The response of high temperature shock and humidity changes prior to and during the early stages of grain development in wheat. Aust J Plant Physiol 17:551–561

Telfer P, Edwards J, Kuchel H et al (2013) Heat stress tolerance of wheat. Australian Grain Technologies (AGT), Grain Research & Development Center, Australia

Tester RF, Morrison WR, Ellis RH et al (1995) Effect of elevated growth temperature and carbondioxide levels on some physic chemical properties of wheat starch. J Cereal Sci 22:63–71

Thomas JMG, Boote KJ, Allen LH Jr et al (2003) Elevated temperature and carbon dioxide effects on soybean seed composition and transcript abundance. Crop Sci 43:1548–1557

Thomson LM (1966) Weather variability, climate change and grain production. Science 188: 535–541

Thorne JH (1982) Characterization of the active sucrose transport system of immature soybean embryos. Plant Physiol 70:953–958

Thurling N (1974) Morpho physiological determinants of yield in rapeseed (*Brassica campestris)* and *(Brassica napus)* II Yield Components. Aust J Agric Res 25:711–721

Thurling N, Vijendra D (1979) Genetic control of the pre anthesis development and seed yield of spring rape (*Brassica napus* L.): ll. Identification of individual genes controlling developmental patterns. Aust J Agric Res 30:261–271

Tian J, Hu X, Gou L et al (2014) Growing degree days is the dominant factor associated with cellulose deposition in cotton fiber. Cellulose 21:813–822

Tollenar M, Daynard TB, Hunter RB (1979) Effect of temperature on rate of leaf appearance and flowering date in maize. Crop Sci 19:363–366

Tomar RKS, Mishra GL (1991) Influence of sowing dates and nitrogen on the yield of mustard (*Brassica juncea* L.). J Oil Seed Res 8:210–214

Torres OA, Palencia E, Lopez de Pratdesaba L et al (2007) Estimated fumonisin exposure in Guatemala is greatest in consumers of lowland maize. J Nutr 137:2723–2729

Triboi-Blondel AMT, Renard M (1999) Effects of temperature and water stress on fatty acid composition of rapeseed oil, 10th International Rapeseed Conference in Canberra, Australia. http://www.regional.org.au/au/gcirc/2/507.htm. Accessed 9 Mar 2016

Uprety DC (2007) Effect of rising atmospheric CO_2 on the productivity of crop plants. In: National fellow report, ICAR, pp 1–26

Uprety DC, Reddy VR (2008) Rising atmospheric CO2 and Crops. ICAR Publications, New Delhi, pp 1–144

Uprety DC, Tomar VK (1993) Photosynthesis and drought resistance of Brassica carinata and its parent species. Photosynthetica 29:321–327
Uprety DC, PrakashS S, Tomar VK (1990) Cytoplasm influences the photosynthetic efficiency in *Brassica carinata*. J Agron Crop Sci 165:261–271
Uprety DC, Chakravarti NVK, Katiyal RK et al (1996) Proceedings of workshop on climate variability of South Asia and its impact on agriculture. Narosa Publication House, New Delhi, pp 262–278
Venkataraman S (1987) Agrometeorological aspects of growth, yield and water relations with special reference to rice. In: Weather and rice. International Rice Research Institute, Philippines, pp 47–54
Venketaswarlu B, Vergara BS, Visperas RM (1987) Influence of photosynthetically active radiation on grain density of rice. Crop Sci 27:1210–1214
Venuprasad R, Dalid CO, Dal Valle M et al (2009) Identification and characterization of large effect quantitative trait loci for grain yield under low land drought stress in rice using bulk segregant analysis (BSA). Theor Appl Genet 120:177–190
Vergara BS, Chang TT, Lilis R (1972) The flowering response of rice plants to photoperiod. IRRI Tech Bull 8:1–55
Vu JCV, Allen LH Jr, Boote KJ et al (1997) Effects of elevated CO_2 and temperature on photosynthesis and rubisco in rice and soybean. Plant Cell Environ 20:68–76
Wahid A, Gilani S, Ashraf M et al (2007) Heat tolerance in plants: an overview. Environ Exp Bot 61:199–223
Walker JM (1969) One degree increments in soil temperature affect maize seedling behaviour. Proc Soc Soil Sci Am 33:729–736
Walton G, Ping S, Bowden B (1999) Environmental impact on canola yield and oil, In: Proceedings of the 10th international rapeseed congress, Canberra, Australia
Wang P, Zhang Z, Song X et al (2014) Temperature variations and rice yields in China: historical contributions and future trends. Climate Change 124:777–789
Wang L, Ma H, Song L et al (2012) Comparative proteomics analysis reveals the mechanism of pre harvest seed deterioration of soybean under high temperature and humidity stress. J Proteomics 75:2109–2127
Wardlaw IF (1974) Temperature control of translocation. In: Bieleske RL et al (eds) Mechanism of regulation of plant growth. Bull Royal Society, Wellington, pp 533–538
Wassman R, Jagdish SVK, Heur S et al (2009) Climate change affecting rice production: the physiological and agronomic basis for possible adaptation strategies. Adv Agron 101:59–122
Wheeler TR, Ellis RH, Hadley P et al (1996a) Assessing the effects of cli-mate change on field crop production. Asp Appl Biol 45:49–54
Wheeler TR, Hong TD, Ellis RH et al (1996b) The duration and rate of grain growth and harvest index of wheat (*Triticum aestivum*) in response to temperature and carbon dioxide. J Exp Bot 47:623–630
White JW, Kimball BA, Wall GW et al (2011) Response of time of anthesis and maturity to sowing dates and infrared warming in spring wheat. Field Crop Res 124:213–222
Whitfield DM (1992) Effect of temperature and ageing on CO_2 exchange of pods of oilseed rape (*Brassica napus*). Field Crop Res 28:271–280
Wilhelm EP, Mullen RE, Keeling PL et al (1999) Heat stress during grain filling in maize: effect on kernel growth and metabolism. Crop Sci 39:1733–1741
Wiseman BR, Isenhour DJ (1989) Effect of temperature on development of corn earworm (Lep:Noct) on meridic diets of resistant and susceptible corn silks. Environ Entomol 18: 683–686
Wittwer SH (1995) Food, climate and CO_2: the global environment and world food production. Michigan State University, Michigan
Wolenweber B, Porter JR, Schllberg J (2003) Lack of interaction between extreme high temperature events at vegetative and reproductive growth stages in wheat. J Agron Crop Sci 189:142–150

Wolf RB, Cavins JF, Kleiman R et al (1982) Effect of temperature on soybean seed constituents: oil, protein, moisture, fatty acids, aminoacids and sugars. J Am Oil Chem Soc 59:230–232

Xu Q, Paulson Q, Guikema JA et al (1995) Functional and ultrastructural injury to the photosynthesis in wheat by high temperature during maturation. Environ Exp Bot 35:43–54

Yamada N, Murata Y, Osada A et al (1955) Photosynthesis of rice plant.II. Proc Crop Sci Soc Jpn 23:112–118

Yamakawa H, Hirose T, Kuroda M et al (2007) Comprehensive expression profiling of rice grain filling- related genes under high temperature using DNA microarray [OA]. Plant Physiol 144:258–277

Yamane Y, Kashino Y, Koike H et al (1998) Effects of high temperatures on the photosynthetic systems in spinach: oxygen evolving activities, fluorescence characterstics and the denaturation process. Photosynth Res 57:51–59

Yaniv Z, Schafferman D, Zur M (1995) The effect of temperature on oil quality and yield parameters of high and low erucic acid cruciferae seeds (rape and mustard). Indian Crop Prod 3:247–251

Yin X, Kropff MJ, Goudriaan J (1996) Differential effects of day and night temperature on development to flowering in rice. Ann Bot 77:203–213

Yin X, Guo W, Spiertz JH (2009) A quantitative approach to characterize sink-source relationships during grain filling in contrasting wheat genotypes. Field Crop Res 114:119–126

Yoon ST, Hoogenboom G, Flitcroft I et al (2009) Growth and development of cotton (Gossypium hirsutum L.) in response to CO_2 enrichment under two different temperature regimes. Environ Exp Bot 67:178–187

Yoshida S (1973) Effects of temperature on growth of the rice plant (*Oryza sativa* L.) in a controlled environment. J Soil Sci Plant Nutr 19(4):299–310

Yoshida S (1978) Tropical climate and its influence on rice. International Rice Research Institute, Philippines

Yoshida S (1981) Fundamentals of rice crop science. International Rice Research Institute, Philippines, p 269

Yoshida S, Parao FT (1976) Climate influence on yield and yield components of lowland rice in the tropics. In: Banos L (ed) Climate and rice. IRRI Publication, Philippines, pp 471–494

Yoshimoto M, Oue H, Kobayashi K (2005) Energy balance and water use efficiency of rice canopies under free air CO_2 enrichment. AgricForest Meteorol 133:226–246

Zahedi M, Sharma R, Jenner CF (2003) Effect of high temperature on grain growth and on the metabolites and enzymes in the starch synthesis pathway in the grains of two wheat cultivars differing in their responses to temperature. Funct Plant Biol 30:291–300

Zakaria S, Matsuda T, Tajima S et al (2002) Effect of high temperature at ripening stage on the reserve accumulation in seed in rice cultivars. Plant Prod Sci 5:160–168

Zamir D (1983) Pollen gene expression and selection: application in plant breeding, part A. Elsevier, Amsterdam, pp 313–330

Zamir D, Tanksley SD, Jones RA (1982) Haploid selection for low temperature tolerance of tomato pollen. Genetics 101:129–137

Zeng B, Xu X, Zhou S (2012) Effects of temperature and light on photosynthetic heterosis of an upland cotton hybrid cultivar. Crop Sci 52:282–291

Zhang S, Tao F (2013) Modeling the response of rice phenology to climate change and variability in different climatic zones: comparison of five models. Eur J Agron 45:165–176

Zhang Y, Tang Q, Peng S et al (2013) Effects of high night temperature on yield and agronomic traits of irrigated rice under field chamber system condition. Aust J Crop Sci 7(1):7

Zhao H, Dai T, Jing Q et al (2007) Leaf senescence and grain filling affected by post anthesis high temperatures in two different wheat cultivars. Plant Growth Regul 51:149–158

Zhao FJ, Su YH, Dunham HJ et al (2009) Variation in mineral micronutrient concentration in grain of wheat lines of diverse origin. J Cereal Sci 49:290–295

Zheng YX, Wu JC, Zhang YP (2012) Photosynthetic temperature responses of four neem (Azadirachta indica A. Juss) provenances in the dry-hot valley area of China. Trees 26(3): 929–937

Zhu XG, Long SP, Ort DR (2010) Improving photosynthetic efficiency for greater yield. Annu Rev Plant Biol 61:235–261

Zhu LJ, Liu QQ, Wilson MH et al (2011) Digestibility and physico chemical properties of rice (*Oryza sativa* L.) flours and starches differing in amylose content. Carbohydr Polym 86: 1751–1759

Ziska LH, Bunce JA (1997) Influence of increasing carbon dioxide concentration on the photosynthetic and growth stimulation of selected C4 crops and weeds. Photosynth Res 54:199–208

Ziska LH, Manalo PA, Ordonez RA (1996) Intraspecific variation in the response of rice (*Oryza sativa*) to increased CO_2 and temperature; growth and yield responses of 17 cultivars. J Exp Bot 47:1353–1359

Ziska LH, Tomecek MB, Gealy DR (2014) Assessment of cultivated & wild, weedy rice lines to concurrent changes in CO_2 concentration and air temperature: determining traits for enhanced seed yield with increasing atmospheric CO_2. Funct Plant Biol 41:236–243

5 Mitigation Technologies to Control High-Temperature Stress in Crop Plants

The temperature of the earth is exponentially rising with an increase by 0.2 °C per decade. It is predicted that the increase in earth's temperature will range between 1.8 and 4.0 °C in 2100 than that of the current value. Rising temperature has become a major concern for crop production. The exponential rise in the concentration of greenhouse gases is considered to be the main cause of global warming (Meehl et al. 2007). The rise in temperature may adversely affect or alter the growth, development, physiological processes, and yield in crop plants. The reaction of crop plants to temperature stress has devastating effect on metabolic processes, disrupting cellular homeostasis and uncoupling major physiological and biochemical processes. The plant responses and adaptations to elevated temperatures and the underlying mechanisms of high-temperature tolerance required to be better understood for important agricultural crops. Plants can adapt to high-temperature stress by various mechanisms of stress avoidance and/or by inherent high-temperature tolerance mechanisms (Fig. 5.1). Among various methods to ameliorate the high-temperature stress effects in plants is presowing seed treatment with low concentrations of inorganic salts, and foliar application of osmoprotectants, signaling molecules, as well as preconditioning of plants is prevalent (Wahid et al 2007). Plants adapt high-temperature stress by altering physiological, morphological, anatomical, and biochemical processes (Fig. 5.2).

The morphological and anatomical alterations include the reduction in cell size, closure of stomata, an increase in stomatal densities, and the greater number of xylem vessels. The biochemical changes to counter the temperature stress effect are to increase proteins, which enhances the activities of antioxidants such as superoxide dismutase, catalase, and peroxidase. These antioxidants scavenge the reactive oxygen species, reduce photooxidation, and maintain the integrity of thylakoid membrane. Similarly, the nutrient management also helps in reducing the adverse effect of temperature changes (Fig. 5.2).

Plants accumulate various metabolites (such as antioxidants, osmoprotectants, heat shock proteins, etc.), and different metabolic pathways of processes are

D.C. Uprety, V.R. Reddy, *Crop Responses to Global Warming*,
DOI 10.1007/978-981-10-2004-9_5

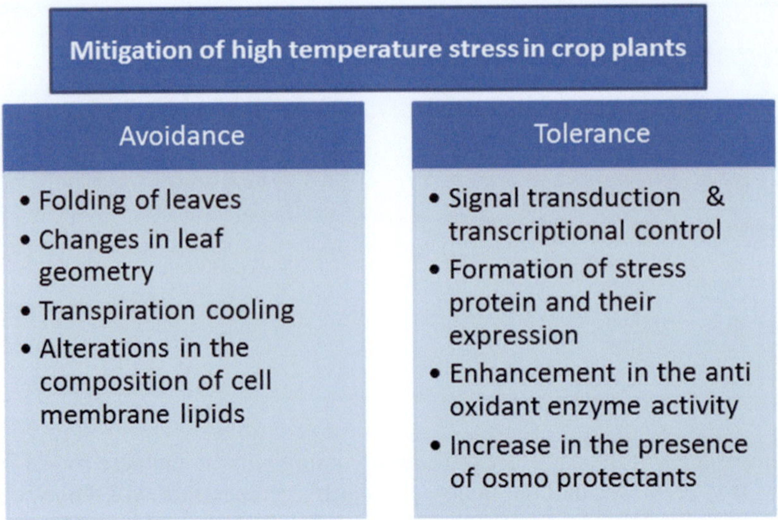

Fig. 5.1 Mitigation of high temperature stress in crop plants

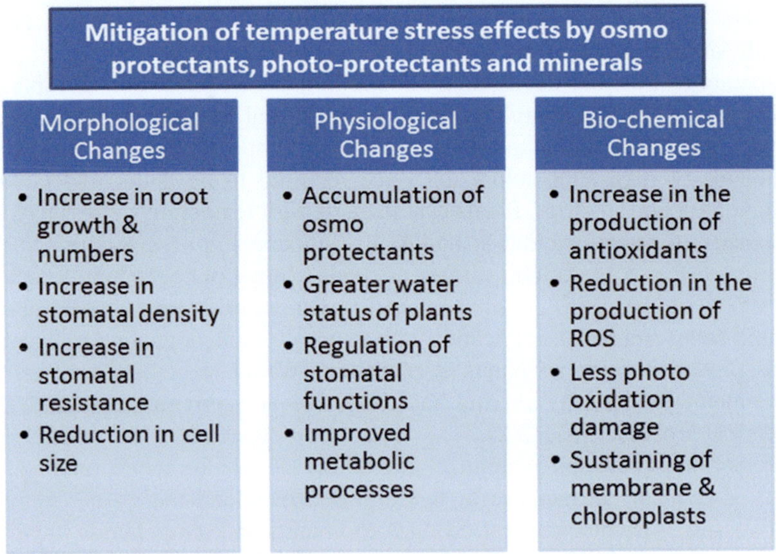

Fig. 5.2 Mitigation of temperature stress by osmo protectants, photo protectants and minerals

activated. The understanding of the signal transduction as well as specific gene expression in response to higher temperature will be important for developing stress-tolerant plants (Zrobek-sokolnik 2012) *and mitigation technologies.*

Survival and sustainability of crop plants to such a rise in temperature will depend on their thermal tolerance. Besides thermal tolerance the avoidance also play a key role in the survival of crop plants. The inherent mechanisms associated with morphological, phenological, and evolutionary adaptations, include changes in leaf orientation, stomatal behavior and their functioning. Early maturing of crops may cause less water loss to escape high-temperature stress (Srivastava et al. 2012). Such morphophysiological adaptations and phenological changes are associated with biochemical processes to save the process of photosynthesis in crop plants. The role of leaf folding under high-temperature stress helps in increasing the efficiency of water metabolism in the flag leaves of wheat under high temperature (Sarieva et al. 2010). Crop management practices, namely, proper sowing method, sowing date, type of cultivars, and methods of irrigation, can also help in avoiding the high-temperature stress. Seed priming by treating seeds with osmotic solutions before sowing helps in countering the temperature stress effects.

Plants have developed various mechanisms to mitigate the adverse effects of high temperature in crop plants. They include leaf orientation, transpiration cooling, and changes in membrane lipid composition (Rodriguez et al. 2005 and Radin et al. 1994). Acclimation mechanisms involved in developing tolerance in crop plants are on transported late embryogenesis abundant (LEA) proteins, osmoprotectants, and antioxidant defense, and factors responsible for signal transduction and their transcriptional control are significant in mitigating the high-temperature stress effects (Wang et al. 2004) & (Rodriguez et al. 2005). The scavenging of ROS (reactive oxygen species) from their site of the generation prior to their diffusion is important for protecting the complex molecules. The main photoprotecting molecules are carotenoids, flavonoids, glutathione, ascorbate, and tocopherols located in the cells and organelles. These photo protectants act (Murchie and Niyogi 2011) by:

1. Avoiding photooxidative damage to membranes and proteins caused by excess light
2. Scavenging ROS like singlet oxygen and dissipate extra energy via xanthophylls mediated nonphotochemical quenching
3. Interacting with membrane lipids, acting like energy receptors and in downregulation process
4. By providing protection through direct quenching or chemically scavenging $^1O_2$, $O_2^{\ -}$, and OH
5. By neutralizing free radicals and ROS before they damage the cells under heat stress
6. By participating in redox signaling

The activities of different antioxidant enzymes have their activation at different temperature ranges. However, the activities of these enzymes increase with the rise in temperature. Chakraborty and Pradhan (2011) observed that catalase, ascorbate peroxidase, and superoxide dismutase showed an initial increase and starts declining at 50 °C, while peroxidase and glutathione reductase activities reduce at temperatures from 20 to 50 °C. The total antioxidant activity was at maximum at 35–40 °C in tolerant and at 30 °C in susceptible cultivars. The susceptibility and tolerance vary in different crop varieties, their growth stages, and growing season (Fig. 5.3).

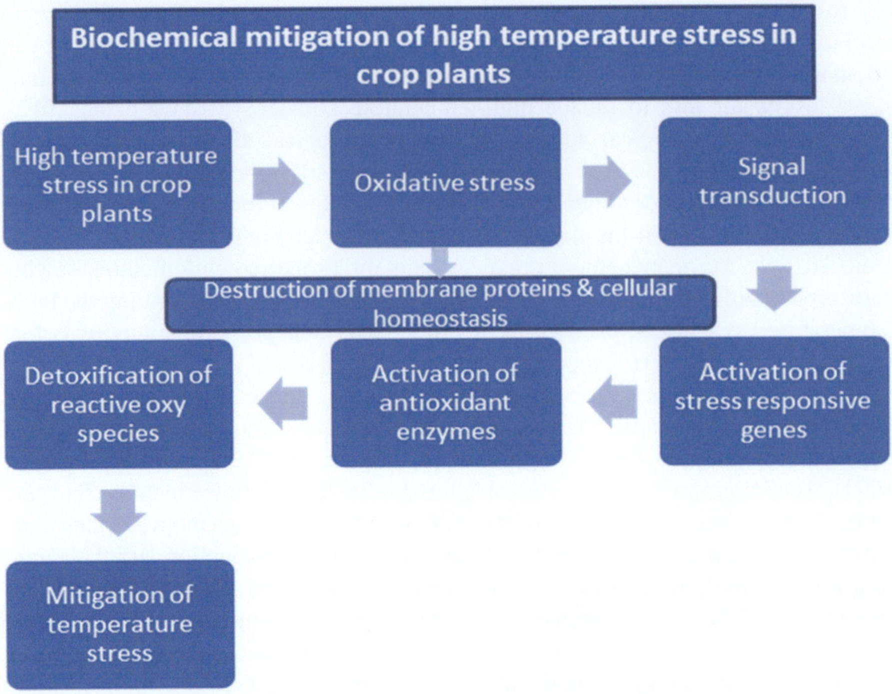

Fig. 5.3 Biochemical mitigation of high temperature stress in crop plants

Ascorbate, glutathione, tocopherol, and carotene protect plants against oxidative stress (Sairam et al. 2000). The high-temperature stress induced accumulation of glutathione and glutathione synthesizing enzymes in wheat (Kocsy et al. 2002). The heat stress increased the glutathione concentration in the flag leaf of two wheat genotypes of contrasting behavior of temperature sensitivity during grain development (Chauhan 2005). Bala et al. (2009) reported that the tolerance of wheat varieties appeared to be correlated with the antioxidant level. Rani et al. (2013) observed that SOD, P ox, CAT, Apx, and GR increased under heat stress, and the increase was greater in tolerant genotypes of *Brassica* compared to susceptible *Brassica* cultivars. Kumar et al. (2012) observed that the expression of enzymatic antioxidants like CAT, APX, and GR was higher in *Zea mays* compared to *Oryza sativa* genotypes. Hasanuzzaman et al. (2013) demonstrated that increase in temperature leads to an increased expression of antioxidant enzymes until a critical temperature in wheat seedlings.

Signal Transduction Once the stress-responsive genes activate, these help to detoxify the ROS (by activating detoxifying enzymes free radical scavengers), to reactivate the essential enzymes and structural proteins (Ciarmiello et al. 2011). These processes help to maintain the cellular homeostasis.

Mitigation of high-temperature stress by exogenous protectants: accumulation of osmolytes such as proline, glycine betaine, and Tre. Supplementation of proline and glycine betaine reduces the H_2O_2 production, improves the accumulation of soluble sugars, and protects the developing tissues from high-temperature stress effects. However, the proline was found more effective than glycine betaine (Rasheed et al. 2011). Kaushal et al. (2011) observed that the application of proline ameliorated the adverse high-temperature effect in chickpea (*Cicer arietinum*) by reducing the injury to the membrane and improved water status and chlorophyll content. Proline application also decreased the MDA and H_2O_2 content by 32 % and 20 %, respectively. Kumar et al (2012) found that exogenous treatment of proline, glycine betaine, and Tre promoted the growth of heat-stressed chickpea plants. Application of protectants such as osmoprotectants, phytohormone signaling molecules, and trace elements shows beneficial effects on plants growing under high-temperature stress. Engineering plants to synthesize these compounds may help in developing thermotolerance in important crop plants and represent a potentially important area of research on thermotolerance.

Exogenous applications of phytohormones were found to be effective in mitigating the high-temperature stress in plants. Salicylic acid is a phytohormone highly effective protectant for plants under heat stress. Dat et al. (1998) reported that the foliar application of salicylic acid in mustard increased the H_2O_2 and reduced the catalase activity, thus increasing their resistance to heat stress. Salicylic acid reduces the electrolyte leakage and catalase activity and increases the glutathione reductase and guaiacol peroxidase activity bringing the resistance against temperature stress. Wang and Li (2006) reported that salicylic acid reduces the thiobarbituric acid reactive substances (TBARS) and reactive electrolyte leakage under heat stress resulting in to high-temperature tolerance in plants. It induces higher Rubisco activation state, accelerating the recovery of net photosynthesis through its effect on PS ll function (Wang et al. 2010). This effect of salicylic acid was related to the enhanced level of heat shock protein (HSP21). The pretreatment of salicylic acid enhances the high-temperature tolerance in crops (Ding et al. 2010) by raising the level of signaling molecules like NO. Stoller et al. (2012) demonstrated that the toxic effect of high temperature on the cells and plant parts (autophagy) can be significantly lowered by the application of cytokinin (0.1 ppm to 3.4 ppm) and potassium (500–4000 ppm) to the leaves and flowers of crop plants. Application of 0.09–0.76 g of cytokinin may be applied per acre along with the aqueous solution of about 0.25–2.0 lbs potassium salts per acre that can induce thermal tolerance in crop plants.

Sustainable agronomic and resource management practices can effectively contribute to temperature stress mitigation for crop plants. At the field level managing or manipulating cultural practices such as timing and methods of sowing, irrigation management, selection of cultivars, and species can also considerably reduce the effect of rising temperature stress on crop plants. For example, early sowing of high-yielding Indian variety of wheat C306 allows longer maturation time with early anthesis that reduces high-temperature exposure to this cultivar, whereas it gave poor yield under late-sown condition (Coiventry et al. 2011) .

Sowing Time Indian farmers adjust sowing time of wheat in such a way so that crop escapes to the hot and desiccating winds during grain-filling period. High-temperature stress during post-flowering stage increases the proportion of gliadins to glutelins and decreases the proportion of starch hampering the quality of wheat flour and the grain yield (Ashraf 2014).

Tillage Zero tillage protects the seedlings from high temperature during initial growth period due to the presence of mulches and crop residues which keeps the soil temperature down during the day and reduces the cooling at night and also helps in conserving the moisture (Geiger et al. 1992). This sustains the metabolic activities in plants without reducing growing period. Rise in soil temperature increased the process of transpiration, which in turn reduces canopy temperature and helps in overcoming the terminal heat stress in plants. Thus senescence reduces and time to maturity delays. According to Tang et al. (2013), straw mulching treatment increases the water holding capacity of soil, thereby mitigating the adverse effect of high temperature on the process of grain filling. Khicher and Niwar (2007) demonstrated that flat bed planting of crops is less vulnerable to high-temperature stress than raised bed planting. The zero tillage, bed planting, and conventional tillage with mulching are significant in mitigating the adverse effect of higher temperature during growth period in wheat crop (Kajla et al. 2015).

Water Management Mitigation of temperature stress is possible by scheduling irrigation according to the critical stages, soil type, environmental conditions, and the quantity of water available for irrigation. The maximum plant height, spikelet m-2, grains spike-1, and thousand grain weight can be obtained at full irrigation and by skipping irrigation during dough and ripening stages as compared to skipping irrigation during seedling, tillering, and booting stages (Tahar et al. 2011). The drip irrigation also helps in maintaining adequate soil moisture, reduces soil temperature to sustain transpiration process, and thereby reduces canopy temperature. Other technologies, such as scheduling of irrigation according to the growth stages of crops, use of efficient irrigation methods, providing extra irrigation if available, and irrigation based on moisture status of soil, help in mitigating the high-temperature stress effects in crop plants.

Foliar Spray High-temperature stress effect is also found to be mitigated by foliar sprays of potassium fertilizer, urea, zinc, IMCP, and GA_3. Field experiments that explore different biochemical, molecular approaches and agronomic management practices are needed to investigate the actual high-temperature responses and their effects on final crop yield.

The application of nutrients such as nitric oxide for signal transduction by detoxifying the high temperature induced ROS, ameliorates the temperature stress effect. (Song et al. 2006). Uchida et al. (2002) reported that nitric oxide induces the expression of gene encoding heat shock protein HSP 26 to protect chloroplast from oxidative damage. Foyer et al. (2002) demonstrated the ameliorating effect of potassium for low-temperature stress by controlling the chilling-induced fluidity of membrane lipids and by improving the photosynthetic electron transport and regulating the stomatal conductance. The calcium mitigates the effect of low-temperature stress by activating the ATPase enzyme of the plasma membrane by recovering the nutrients lost by cold-induced cell damage (Palta 2000). Rawal et al. (2007) demonstrated that calcium treatment to potato plants mitigates the adverse heat stress effect. They attributed it to improved cell membrane stability. Nissen and Orcutt (1996) suggested that calcium has a condensing effect on the phospholipids in the bilayer making them rigid and strong, thereby protecting the cell membrane from high-temperature stress. Calcium in the form of calmodulin mitigates the low-temperature stress effects on plant metabolic processes and plant growth. Waraich et al. (2012)) reported that magnesium improves the root growth, increases the water and nutrient uptake, and significantly decreases the temperature stress-induced ROS generation.

The temperature-induced adverse effects on plants include protein aggregation and denaturation, enhanced membrane lipid fluidity, and damage of membrane integrity. Physiological, biochemical, morphological, and anatomical adaptation measures and agronomic management help in mitigating the temperature stress effects in crop plants.

Application of nutrients like nitrogen, potassium, calcium, and magnesium decreases the damaging effects of reactive oxygen species on the membrane lipids by increasing the effect of antioxidant enzymes such as SOD, CAT, and POD in the cells of plant (Almeselmani et al. 2006). Application of potassium and calcium mitigates the high-temperature effect by maintaining the water status of plants under high-temperature condition through osmotic adjustment and regulation of stomatal functions. The salicylic acid and selenium treatments reduce the membrane damage by increasing the antioxidant enzyme activities.

The changes in cultural practices such as sowing time, method of sowing, selection of cultivars, and irrigation management significantly reduce the adverse effect of temperature stress in crop plants. Application of osmoprotectants, phyto-hormones, signaling molecules, and trace elements ameliorates the adverse effect of high-temperature stress in crop plants. Engineering plants, to synthesize these compounds, may help in developing high-temperature-tolerant plant and is an important area for research on thermotolerance (Awasthi et al. 2015, Ashraf et al. 1994). Field experiments to investigate various biochemical and molecular approaches and agronomic management practices are required to be studied to understand the high-temperature responses and their effects on crop yield.

References

Almeselmani M, Deshmukh PS, Sairam RK, Kushwah SR, Singh TP (2006) Protective role of anti oxidant enzymes under high temperature stress. Plant Sci 171:382–388

Ashraf M (2014) Stress induced changes in wheat grain composition and quality. Crit Rev Food Sci Nutr 54:1576–1583

Ashraf M, Saeed MM, Qureshi MJ (1994) Tolerance to high temperature in cotton (Gossypium hirsutum L.) at initial growth stages. Environ Expt Bot 34:275–283

Awasthi R, Bhandari K, Nayyar H (2015) Temperature stress and redox homeostasis in agricultural crops. Front EnvSci 3:1

Balla K, Bencze S, Janda T, Veisz O (2009) Analysis of heat stress tolerance in winter wheat. Acta Agron Hunga 57:437–444

Chakraborty U, Pradhan D (2011) High temperature induced oxidative stress in *Lens culinaris*. Role of anti oxidants and amelioration of stress by chemical pretreatments. J Plant Interact 6:43–52

Chauhan S (2005) Ph.D. thesis HNB Garhwal University. Physiological and molecular basis of heat tolerance with emphasis on oxidative stress metabolism in wheat

Ciarmiello LF, Woodrow P, Fuggi A, Pontecorvo G, Carillo P (2011) Plant genes for abiotic stress. In: Venketswar P, Shanker AK (eds) Abiotic stress in plants- mechanisms and adaptations. In Tech, Rijecka, pp 283–308

Coventry DR, Gupta RK, Yadav RS, Poswal RS, Chhokar RK, Sharma VK, Yadav SC, Gill A, Kumar P, Mehta SG, Kleeman A, Bonamano A, Cumins JA (2011) Wheat quality and productivity as affected by varieties and sowing time in Haryana. India Field Crop Res 123: 214–225

Dat JF, Lopez-Delgado H, Foyer CH, Scott IM (1998) Parallel changes in H_2O_2 and catalase during thermotolerance induced by salicylic acid or heat acclimation in mustard seedlings. Plant Physiol 116:1351–1357

Ding W, Song L, Wang X, Bi Y (2010) Effect of abscisic acid on heat stress tolerance in the Calli from two ecotypes of *Phragmites communis*. Biol Planta 54:607–613

Foyer CH, Vanacker H, Gornez LD, Harbinson J (2002) Regulation of photosynthesis and antioxidant metabolism in maize leaves at optimum and chilling temperatures. Review. Plant Physiol Biochem 40:659–688

Geiger SC, Manu A, Balviano A (1992) Changes in a sandy Sahelian soil following crop residue and fertilizer additions. Soil Sci Soc Am J 56:172–177

Hasanuzzaman M, Nahar K, Alam MM, Roychowdhury R, Fujita M (2013) Physiological, biochemical and molecular mechanisms of heat stress tolerance in plants. Int J Mol Sci 14: 9643–9684

Kajla M, Yadav VK, Chhokar RS, Sharma RK (2015) Management practices to mitigate the impact of high temperature on wheat. J Wheat Res 7:1–12

Kaushal N, Gupta K, Bhandari K, Kumar S, Thakur P, Nayyar H (2011) Proline induces heat tolerance in chickpea (Cicer arietinum L) plants by protecting vital enzymes of carbon & antioxidation metabolism. Physiol Mol Biol Plants 17:203–213

Khicher ML, Niwar R (2007) Thermal effect on growth and yield of wheat under different soil environment and plant systems. Indian J Agric Res 41:92–96

Kocsy G, Szali G, Galiba G (2002) Effect of heat stress on glutathione biosynthesis in wheat. Acta Biol Szeged 46:71–72

Kumar S, Gupta D, Nayyar H (2012) Comparative response of maize and rice genotypes to heat stress. Status of oxidative stress and anti oxidants. Acta Physiol Plant 34:75–86

Meehl GA et al (2007) Climate change 2007. Cambridge University Press, Cambridge, pp 749–845

Murchie EH, Niyogi KK (2011) Manipulation of photoprotection to improve plant photosynthesis. Plant Physiol 155:86–92

Nissen TE, Orcutt DM (1996) Plant membranes as environmental stresses. In: Physiology of plants under abiotic stress factors. Willey, New York, pp 66–68

Palta JP (2000) Stress interactions at cellular and membrane levels. Hortic Sci 25:1377
Radin JW, Lu Z, Percy RG, Zeiger E (1994) Genetic variability of stomatal conductance in Prima cotton and its relation to improvement of heat adaptation. Proc Natl Acad Sci U S A 91: 7217–7221
Rani B, Dhawan K, Jain V, Chopra ML, Singh D (2013) High temperature induced changes in anti oxidative enzymes in Brassica juncea L. Czern Coss. Available online http://www.australianoilseeds.com/pdffile0003/6861/96
Rasheed R, Wahid A, Farooqi M, Hussain I, Basra SMA (2011) Role of proline and glycine betaine pretreatment in improving heat tolerance of sprouting sugarcane buds (Saccharum, sp.). Plant Growth Regul 65:35–45
Rawal S, Rana NS, Kumar D (2007) Mitigation of heat stress in potato through calcium nutrition. Potato J 34:111–112
Rodriguez M, Canales E, Borras Hidalgo O (2005) Molecular aspects of abiotic stress in plants. Biotechnol Appl 22:1–10
Sairam RK, Srivastava GC, Saxena DC (2000) Increased antioxidant activity under elevated temperature: a mechanism of heat stress tolerance in wheat genotypes. Biol Planta 43:245–251
Sarieva GE, Kenzhebaeva SS, Lichtentheler HK (2010) Adaptation potential of photosynthesis in wheat cultivars with a capability of leaf rolling under high temperature conditions. Russ J Plant Physiol 57:28–36
Song L, Ding W, Zhao M et al (2006) Nitric oxide protects against oxidative stress under heat stress in the calluses from two ecotypes of reed. Plant Sci 171:449–458
Srivastava S, Pathak AD, Gupta PS, Srivastava AK, Srivastava AK (2012) Hydrogen peroxide scavenging enzymes impart tolerance to high temperature induced oxidative stress in sugarcane. J Environ Biol 33:657–661
Stoller J, Liptay A, Salzman R (2012) Composition and methods for stress mitigation in plants. Publication no.US2012/0252673. A 1
Tahar AL, Ali B, Hwary EL, Samia OY (2011) Effect of skipping irrigation on growth, yield components and water use efficiency of wheat in semi arid region of Sudan. Agric Biol J North Am 60:1003–1009
Tang Y, Xiolin W, Chiaosk L, Chun W, Xiaoling M, Gang H (2013) Long term effect of year round tillage patterns on yield and grain quality of wheat. Plant Prot Sci 16:365–373
Uchida AT, Jagendorf T, Hibino (2002) Effect of hydrogen peroxide and nitric oxide on both salt and heat stress tolerance of rice. Plant Sci 163:515–523
Wahid A, Gilani S, Ashraf M, Foolad MR (2007) Review heat tolerance in plants: an overview. Environ Exp Bot 31:199–223
Wang W, Vinocur B, Shoreyov O, Altman A (2004) Role of plant heat shock proteins and molecular chaperons in the abiotic stress response. Trends Plant Sci 9:244–252
Wang LJ, Li SH (2006) Salicylic acid induced heat or cold tolerance in relation Ca 2+ homeostasis and anti oxidant systems in young grape plants. Plant Sci 170:685–694
Wang LJ, Fan L, Loeschar W, Duan W, Liee GJ, Chang JS, Luo HB, Li SH (2010) Salicylic acid alleviates decreases in photosynthesis under heat stress and accelerates recovery in grape wine leaves. BMC Plant Biol 10:34
Waraich EA, Ahmad R, Halim A, Aziz T (2012) Alleviation of temperature stress by nutrient management in crop plants- a review. J Soil Sci Plant Nutr 12:221–244
Zrobek-Sokolink A (2012) The temperature stress and response of plants. In: Mahmood G, Prasad MNV (eds) Environmental adaptations and stress tolerance in plants in the era of climate change. Springer, New York, pp 113–134

MIX
Papier aus verantwortungsvollen Quellen
Paper from responsible sources
FSC® C105338

If you have any concerns about our products,
you can contact us on
ProductSafety@springernature.com

In case Publisher is established outside the EU,
the EU authorized representative is:
Springer Nature Customer Service Center GmbH
Europaplatz 3, 69115 Heidelberg, Germany

Printed by Libri Plureos GmbH
in Hamburg, Germany